Finite Element
Programming

COMPUTATIONAL MATHEMATICS AND APPLICATIONS

Series Editor
J. R. WHITEMAN

Institute of Computational Mathematics, Brunel University, England

E. HINTON and D. R. J. OWEN: Finite Element Programming

M. A. JASWON and G. T. SYMM: Integral Equation Methods in Potential Theory and Elastostatics

Finite Element Programming

E. HINTON
and

D. R. J. OWEN

Department of Civil Engineering
University College of Swansea, U.K.

ACADEMIC PRESS, INC.

(Harcourt Brace Jovanovich, Publishers)

Orlando   San Diego   New York   London
Toronto   Montreal   Sydney   Tokyo

ACADEMIC PRESS INC. (LONDON) LTD.
24/28 Oval Road,
London NW1

United States Edition published by
ACADEMIC PRESS, INC.
Orlando, Florida 32887

Library of Congress Catalog Card Number: 77 77366

ISBN Hardback: 0-12-349350-1
ISBN Paperback: 0-12-349352-8

PRINTED IN THE UNITED STATES OF AMERICA

85 86 87 88   9 8 7 6 5

Editor's Foreword

This new series on computational mathematics has been conceived in order to fill the gap which exists between numerical mathematics and theoretical applied mathematics on the one hand and engineering and scientific applications of numerical methods on the other. The numerical theme is central to the series and thus topics suitable for inclusion will range from numerical analysis to the application of numerical methods in engineering. The series will consist of text books, monographs and conference proceedings spanning computational mathematics and its applications. Naturally some books will be so specialised as to be limited either to theory or to practice. However, the emphasis throughout will be on a readable presentation and it is intended that all the books in the series will be tools from which persons can readily learn.

It is fitting that Book I of the series should be on the subject of finite elements. The finite element method is now well established as an engineering tool with wide application. At the same time it has attracted considerable attention from mathematicians over the last ten years, so that a large body of mathematical theory now exists. Finite elements thus span theory and practice and fall admirably into the spirit of the series. One great difficulty that the would-be user of finite elements experiences is the apparent complexity of programming the method right from the start. It is now widely acknowledged that the programming involved in implementing finite elements is considerably more complicated than that of rival methods such as finite differences. This has deterred people from moving into the finite element field. Drs Hinton and Owen have gone a long way towards explaining the problems and pitfalls of finite element programming and their clear presentation will help students in their own applications. The authors write in the light of many years experience in teaching finite elements to civil engineering students at Swansea and in research on the subject.

It is thought that the series in general and this book in particular will prove important additions to the literature.

Brunel University   J. R. Whiteman
January 1977

Preface

The finite element method has attracted a wide variety of theoreticians and practitioners from various disciplines, including engineering, mathematics and computer science. The finite element textbooks currently available emphasise the basic theoretical aspects of the method with applications being presented to demonstrate the essentially practical nature of the technique. However, as anyone who has ever tried to develop a finite element program will testify, there is an enormous gulf between the basic theory and a working computer code. It is true that some finite element textbooks have presented, almost as an afterthought, a finite element computer program: usually included as an appendix. No textbook appears to have been devoted primarily to the programming aspects of the technique.

With these thoughts in mind, we decided to write a book describing in detail three specially written finite element programs. The basic aim was that these programs should help the reader to take the painful step from theory to program, thus enabling him to develop (or at least appreciate) programs for his own particular applications in his own environment. In the development of the programs, the accent has been placed on simplicity, ease of understanding and practicality. Only one type of element is used throughout—namely the *parabolic isoparametric element*. This curved-sided, well tested, good performer allows us to demonstrate all the features of a finite element computer code and, in particular, the role of numerical integration. Equation solving is performed using the powerful frontal solution technique.

This book owes much to the remarkable work of Bruce Irons which laid the basis, not only of much analytical work on the finite element method, but also the programing techniques necessary to make the method a practical tool. In particular, we acknowledge Professor Irons as the originator of the frontal solution process and the parabolic isoparametic element. He also drew our attention to the importance of error diagnostics in finite element programs.

We would like to take this opportunity to thank others for their direct or indirect assistance: Professor O. C. Zienkiewicz who, over the years, has greatly stimulated our interest in the finite element method, Dr G. A. Fonder who helped us to gain a new perspective on the frontal solution process and contributed considerably to its description in Chapter 8, Professor J. R. Whiteman, the editor of this series, who made many valuable suggestions, comments and criticisms, all our colleagues and research students on whom the ideas presented in this text have been tested, and finally Mrs M. J. Davies for her meticulous care in typing the manuscript, Mr R. Edwards for preparing the diagrams and drawings and Mr P. Ellison for his advice on the production of the program listings.

Swansea
January 1977

E. Hinton
D. R. J. Owen

Contents

To our Wives

1

Introduction and Theory

1.1 Introduction

The finite element method is now firmly established as an engineering tool of wide applicability. No longer is it regarded as the sole province of the researcher or academic but it is now employed for design purposes in many branches of technology. One of the principal advantages of the finite element method is the unifying approach it offers to the solution of diverse engineering problems.

During its early development for stress analysis problems the method relied heavily on a physical interpretation in which the structure was assumed to be composed of elements physically connected only at a number of discrete nodal points. Later the application of the method to structural mechanics problems was developed through the use of the principle of virtual work and energy methods. The method was then generalised and its wider mathematical roots were recognised; it was shown that finite elements could be applied to any mathematical problem for which a variational functional existed. More recently, finite element solutions have been developed which are based on the well known, classical techniques known as "weighted residual methods", e.g. Galerkin, collocation and least squares approximation. In fact the method is now widely recognised as a general numerical technique, for the solution of partial differential equation systems subject to appropriate boundary and initial conditions.

In engineering, physics and applied mathematics, three main areas of application of the finite element method can be identified [1–4]. These are:

Equilibrium problems in which the system does not vary with time. Examples of such problems include the stress analysis of linear elastic systems, electrostatics, magnetostatics, steady-state thermal conduction and fluid flow in

1

porous media. The structure is first divided into distinct non-overlapping regions known as *elements* over which the main variables are interpolated. These elements are connected at a discrete number of points along their periphery known as *nodal points*. Fig. 1.1 shows the finite element model used in the linear elastic stress analysis of an industrial fan, where it is required to find the displacements and stress components.

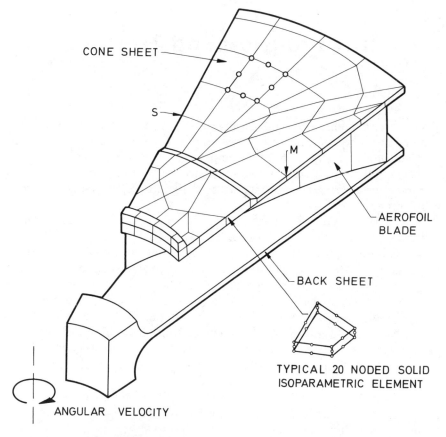

CONE SHEET

S

M

AEROFOIL
BLADE

BACK SHEET

TYPICAL 20 NODED SOLID
ISOPARAMETRIC ELEMENT

ANGULAR VELOCITY

FIG. 1.1. Repeatable section of a sectorially symmetric three-dimensional fan.

This three dimensional problem was solved using *twenty noded isoparametric brick elements* and its complex geometrical configuration is typical of the type of problem for which the finite element method is indispensable.

Eigenvalue problems are extensions of equilibrium problems in which specific or critical values of certain parameters must be determined. The stability of structures and the determination of the natural frequencies of linear elastic systems are examples of such problems. In a finite element solution of a vibration problem, each mode shape or eigenvector is associated with a particular frequency or eigenvalue. Fig. 1.2 shows the modes of vibration of a cantilever plate, [5], which were calculated using four triangular plate bending elements.

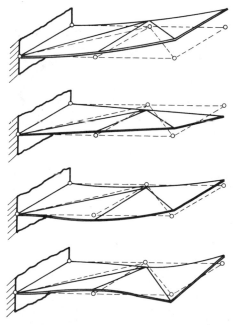

FIG. 1.2.   Vibration of a cantilever plate divided into four triangular elements: modal shapes.

Propagation problems include problems in which some time-dependent phenomena takes place. Hydrodynamics and the dynamic transient analysis of elastic continua are two examples of such problems. Fig. 1.3 gives the results of a finite element hydrodynamic analysis of the North Sea, [6], in which approximation to the current velocity components are sought. A tidal variation is specified by prescribing the height above mean sea level and compatible velocities as sinusoidal oscillations on appropriate boundaries. Inflows from the rivers Thames and Maas are also specified and Fig. 1.3 shows the current velocities plotted as vectors at a particular time.

In each of these three areas of application, problems may contain some non-linear characteristic which complicates the analysis. A typical non-

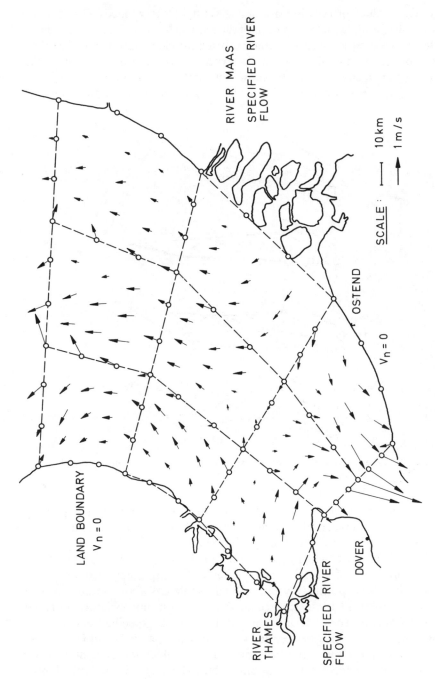

FIG. 1.3.   North Sea current vector plot.

linear problem is shown in Fig. 1.4 which illustrates the elasto-plastic stress analysis of a spherical pressure vessel with a flush nozzle junction, [7]. The geometry of the vessel is shown in Fig. 1.4 and indicates the weld fillet angle considered. The development of zones of plastic yielding with increasing pressure are also shown.

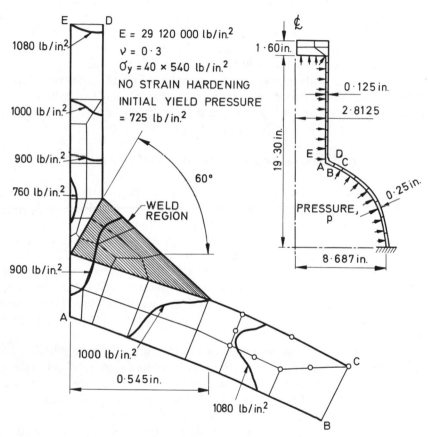

FIG. 1.4. Finite element idealisation by isoparametric elements of a spherical steel pressure vessel with a flush nozzle junction showing plastic zones developed.

The success of the finite element method as a practical design aid depends on the availability of an efficient means of solving the resulting system of linear or non-linear simultaneous equations. Clearly the existence of the computer is vital to the success of this.

Increases in core storage capacities to the present level have allowed a wide variety of problems to be comfortably processed without the need for sophisticated data handling techniques. However the parallel development of auxiliary hardware, such as direct access discs to replace magnetic tape systems, has permitted the solution of a new magnitude of problem. For example, it is not uncommon in the design of off-shore structures to analyse systems containing more than 10,000 degrees of freedom. This continuing trend is also revolutionising the position of non-linear applications.

In the past, the economic limitations imposed by computer costs have restricted the general use of such techniques. However this barrier is being rapidly removed and the finite element solution of such problems is already economically acceptable for selected industrial applications.

Such developments, along with future enhancements in element characteristics, equation solution techniques, etc., suggest that the finite element method will play a major role in engineering design for many years to come.

1.2 General procedures and discretisation by finite elements

In this text we limit our attention to the application of the finite element method to linear equilibrium problems. Furthermore, detailed consideration will be given only to the finite element displacement method of structural analysis. Nonetheless, the basic similarities which exist between the various linear equilibrium problems is reflected in the corresponding finite element models. Consequently much can be learnt from the particular equilibrium model which we have adopted as our basis. Although eigenvalue and propagation problems are not specifically considered, it is hoped that those readers interested in solving such problems by finite elements will still find this text useful.

In order to define terms we review briefly the finite element method. In any continuum the actual number of degrees of freedom is infinite and, unless a closed form solution is available, an exact analysis (within the assumptions made) is impossible. For any numerical approach an approximate solution is attempted by assuming that the behaviour of the continuum can be represented by a finite number of unknowns. As previously mentioned in the finite element method the continuum is divided into a series of elements which are connected at a finite number of points known as nodal points. This process is known as discretisation and a typical subdivision is shown, for example in Fig. 1.4.

For structural applications at least, the governing equilibrium equations can be obtained by minimising the total potential energy of the system. The total potential energy, π, can be expressed as

$$\pi = \tfrac{1}{2} \int_V [\boldsymbol{\sigma}]^T \boldsymbol{\varepsilon} \, dV - \int_V [\boldsymbol{\delta}]^T \mathbf{p} \, dV - \int_S [\boldsymbol{\delta}]^T \mathbf{q} \, dS, \qquad (1.1)$$

where $\boldsymbol{\sigma}$ and $\boldsymbol{\varepsilon}$ are the *stress* and *strain* vectors respectively, $\boldsymbol{\delta}$ the *displacements* at any point, $\mathbf{p}$ the *body forces* per unit volume and $\mathbf{q}$ the *applied surface tractions*. Integrations are taken over the volume V of the structure and loaded surface area, S.

The first term on the right hand side of (1.1) represents the internal strain energy and the second and third terms are respectively the work contributions of the body forces and distributed surface loads.

In the finite element displacement method, the displacement is assumed to have unknown values only at the nodal points, so that the variation within any element is described in terms of the *nodal values* by means of interpolation functions. Thus

$$\boldsymbol{\delta} = \mathbf{N}\boldsymbol{\delta}^e, \tag{1.2}$$

where $\mathbf{N}$ is the set of interpolation functions termed the *shape functions* and $\boldsymbol{\delta}^e$ is the vector of nodal displacements of the element. The strains within the element can be expressed in terms of the element nodal displacements as

$$\boldsymbol{\varepsilon} = \mathbf{B}\boldsymbol{\delta}^e, \tag{1.3}$$

where $\mathbf{B}$ is the *strain matrix* generally composed of derivatives of the shape functions. Finally the stresses may be related to the strains by use of an *elasticity matrix* $\mathbf{D}$, as follows

$$\boldsymbol{\sigma} = \mathbf{D}\boldsymbol{\varepsilon}. \tag{1.4}$$

Provided that the element shape functions have been chosen so that no singularities exist in the integrands of the functional, the total potential energy of the continuum will be the sum of the energy contributions of the individual elements. Thus

$$\pi = \sum_e \pi_e, \tag{1.5}$$

where π_e represents the total potential energy of element e which, on use of (1.1), can be written

$$\pi_e = \tfrac{1}{2} \int_{V_e} [\boldsymbol{\delta}^e]^{\mathrm{T}} [\mathbf{B}]^{\mathrm{T}} \mathbf{D} \mathbf{B} \boldsymbol{\delta}^e \, dV - \int_{V_e} [\boldsymbol{\delta}^e]^{\mathrm{T}} [\mathbf{N}]^{\mathrm{T}} \mathbf{p} \, dV - \int_{S_e} [\boldsymbol{\delta}^e]^{\mathrm{T}} [\mathbf{N}]^{\mathrm{T}} \mathbf{q} \, dS, \tag{1.6}$$

where V_e is the element volume and S_e the loaded element surface area. Performance of the minimisation for element e with respect to the nodal displacements $\boldsymbol{\delta}^e$ for the element results in

$$\frac{\partial \pi_e}{\partial \boldsymbol{\delta}^e} = \int_{V_e} ([\mathbf{B}]^{\mathrm{T}} \mathbf{D} \mathbf{B}) \boldsymbol{\delta}^e \, dV - \int_{V_e} [\mathbf{N}]^{\mathrm{T}} \mathbf{p} \, dV - \int_{S_e} [\mathbf{N}]^{\mathrm{T}} \mathbf{q} \, dS,$$

$$= \mathbf{K}^e \boldsymbol{\delta}^e - \mathbf{F}^e, \tag{1.7}$$

where

$$\mathbf{F}^e = \int_{V_e} [\mathbf{N}]^T \mathbf{p} \, dV + \int_{S_e} [\mathbf{N}]^T \mathbf{q} \, dS, \qquad (1.8)$$

are the *equivalent nodal forces* for the element, and

$$\mathbf{K}^e = \int_{V_e} [\mathbf{B}]^T \mathbf{D} \mathbf{B} \, dV, \qquad (1.9)$$

is termed the *element stiffness matrix*. The summation of the terms in (1.7) over all the elements, when equated to zero, results in a system of equilibrium equations for the complete continuum. These equations are then solved by any standard technique to yield the nodal displacements.

Assembly is covered in greater detail in Chapter 2. The stresses within each element can then be calculated from the displacements using (1.3) and (1.4).

The basic steps for deriving a finite element solution to an equilibrium problem can be summarised as

Sub-division of the continuum into finite elements.

Evaluation of element *stiffness* and *load* terms.

Assembly of element *stiffness* and *load* terms into an overall *stiffness* matrix and *load* vector.

Solution of the resulting linear simultaneous equations for the unknown nodal variables.

Evaluation of subsidiary element quantities such as stresses in the displacement method.

1.3 The basic aims and scope of the book

This text is concerned with the programming of the finite element method. There is a tremendous difference between the understanding of the theory behind the finite element process and the ability to program the method for practical application. It is the object of this book to try to bridge this gap.

It is hoped that the book will be useful in a teaching as well as a research and development environment and that undergraduates, research workers and practising engineers alike will benefit from this self-contained text, which is intended to serve as both a reference book for finite element courses and to assist students in taking the painful step from theory to program.

It is appreciated that potential readers will have widely differing backgrounds and previous levels of finite element experience. Since the main aim of the book is to indicate how the major transition from theory to program is taken, all programming has been deliberately made as transparent and uncomplicated as possible. Economy has been sacrificed for transparency where any conflict arises. At all stages the authors have asked the question "How will this section help the reader to make a smooth transition from theory to program?" At the same time an attempt has been made to provide programs which are useful in a research environment (particularly after the enhancements discussed in Chapter 11 have been incorporated) and which the more sophisticated reader can readily extend and develop.

It is however important to emphasise that the programs developed in this book should only be used when the reader is satisfied that he understands them completely.

A large proportion of time spent in any finite element analysis is that required for checking the input data and correcting the errors which inevitably occur. Thus appreciable savings in both computation time and man-hours expended accrue if the input data can be scrutinised and any errors detected in some automatic manner before computation begins. For this purpose, error diagnostic subroutines are included in the programs presented. These check the input data and if any errors are detected, appropriate diagnostic messages are printed.

However, the reader should be warned at the outset that this does not give a license to careless data preparation, since error diagnostic subroutines, however sophisticated, cannot guarantee to detect all data errors.

As mentioned in the previous section, only structural problems are considered in the text; however upon completion of the book the reader should be in a position to produce programs for a variety of applications. In this sense the text can be treated as a detailed Program and User's Guide in much the same way as a Workshop Manual would be employed when maintaining a motor car.

In order to program the finite element method effectively the programmer must:

Have a working knowledge of some programming language. In this text all programs are written in FORTRAN.

Be familiar with matrix methods. Throughout any program, operations such as matrix multiplication, inversion, etc., occur frequently. Indeed prior experience with the matrix analysis of structural frameworks serves as a useful introduction to the finite element processes.

Have a knowledge of finite element methods. A complete familiarity with the theoretical expressions of the technique is, of course, essential.

Borrow some tools from numerical analysis. For example the concept of interpolation is required and, if complicated elements are to be employed, the use of numerical integration is unavoidable. However by far the most important numerical aspect to be considered must be the solution of the resulting system of simultaneous equations. In this text a Gaussian elimination technique is adopted.

Be aware of the physics and engineering assumptions of the problem.

When writing finite element programs, one has to choose between developing an integral program with very few subroutines or producing a modular program in which different operations are carried out in individual subroutines. As seen in the previous section, the basic theoretical steps of the finite element process can be separated into several distinct phases. Therefore it is logical to write finite element programs in which each theoretical operation is performed in a separate subroutine and such an approach is adopted here. The structure of the programs developed is described in detail in Section 1.6.

The layout of the book is such that as soon as a section of theory is completed the programming concerned with it is undertaken. In this way a series of modular subroutines are initially presented which are later assembled to form the complete program. All difficult programming problems are discussed in detail.

The applications considered in this text are:

(a) Beam analysis

(b) Plane stress and strain situations

(c) Plate bending problems.

For each application the parabolic isoparametric element is employed. All associated stiffness matrices and load vectors are numerically integrated.

In Section 1.5 the basic expressions of the finite element method for structural applications are presented. Some basic methods of simultaneous equation solution are reviewed in Chapter 2, with an emphasis placed on Gaussian elimination techniques. At this stage some FORTRAN programming is introduced, and simple subroutines for equation solution are developed. Chapter 3 deals with the problems associated with input and output. The input data required for finite element analysis with isoparametric elements is discussed and subroutines for data assimilation are presented.

Chapter 4 is devoted entirely to the development of specific expressions and subroutines for the isoparametric beam element. It is at this stage that problems specifically associated with the isoparametric element concept are first encountered and the beam element serves as a convenient introductory vehicle in view of its relative simplicity. In particular the Jacobian matrix, which enables transformation between local and global quantities to be made, is introduced and numerical integration techniques essential to isoparametric elements are discussed. The subroutines performing the standard steps, such as shape function and stiffness formulation, equivalent nodal force generation and stress resultant evaluation, are developed at this stage. These subroutines are essentially of the same form as for more sophisticated applications and Chapter 4 therefore allows the reader to familiarise himself with the general structure of isoparametric element programs.

Chapters 5 and 6 present the basic expressions and subroutines for plane stress/strain and plate bending applications respectively. In particular all the subroutines necessary for the construction of the element stiffness matrices and for the evaluation of the stresses from the known displacements are developed.

In the finite element method all loading must be applied to the structure as equivalent nodal loads. Since several forms of loading such as gravity, pressure, thermal effects, may be applied to a structure, subroutines are necessary to compute the nodal force equivalents and Chapter 7 is devoted to the development of such subroutines.

The solution of equation systems by the frontal method is dealt with in Chapter 8 and a sophisticated subroutine is developed which can be employed as a general purpose finite element solver. The frontal equation solution technique is described in detail and its advantages outlined.

In Chapter 9 three subroutines are presented for input data checking. The data is checked in stages and if any errors are detected, appropriate diagnostic messages are printed and the remainder of the input data is echoed by line-printer.

The subroutines developed in Chapters 3 to 9 are assembled in Chapter 10 to form complete programs which can be employed for beam analysis, plane stress/strain problems or plate bending situations. Numerical examples for each application are also presented demonstrating the efficiency of the parabolic isoparametric element.

In the final chapter the various programs presented are discussed. Extensions to other situations are suggested and areas in which improvements and modifications can be made are indicated.

1.4 The use of parabolic isoparametric elements

A single type of element is used throughout this book; namely the parabolic isoparametric element, a typical two-dimensional version of which is illustrated in Fig. 1.5. This is a deliberate policy, since a text book filled with a variety of different elements would tend to confuse and distract the reader. Instead,

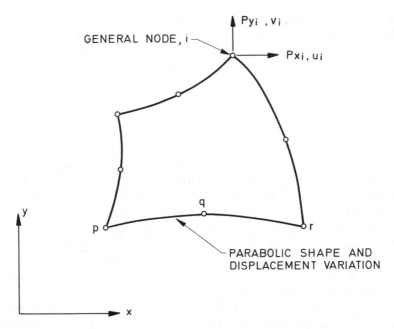

FIG. 1.5.   Typical two-dimensional parabolic isoparametric element.

for a minimum amount of effort, the reader can fully understand the principles behind the formulation of the parabolic isoparametric element and then reap the benefits from the many applications which follow almost automatically. Parabolic isoparametric elements are extremely versatile, good performers and are well tried and tested. Practical experience suggests that, for a given number of total degrees of freedom in a structure, greater accuracy is achieved by use of fewer complex elements in place of a larger number of simple elements. This fact is substantiated by theoretical error analysis provided the solution is sufficiently smooth. In non-linear problems the benefits can be even more marked than for the linear case. After he has seen how parabolic isoparametric elements fit into a program, the reader should then be in a position to insert other elements of his own choice.

1.5 The finite element displacement method

1.5.1 Basic Nodal Variables—generalised displacements

Although this text considers only structural applications several options still remain open for the choice of nodal variables and problems can be formulated in several different ways. Alternative approaches are offered by considering either the displacements or the stresses or a combination of both as the basic variables. If the displacements are chosen as the prime unknowns, with the stresses being determined from the calculated displacement field, the process is termed *the displacement method* and appeals to engineers in view of its similarity to the displacement method of matrix analysis. The principal steps of this procedure have been already outlined in Section 1.2.

Alternatively it is possible to proceed with the stresses as the prime unknowns; an approach which is termed *the equilibrium method*. If both stresses and displacements are employed as variables simultaneously the method is said to be *mixed* or *hybrid*. In this book only displacement models are employed.

The displacement variables used are termed *generalised displacements* since the components involved may not all be physically interpreted directly as displacements. For example, Fig. 1.6 indicates the situation in plate bending analysis. Here, at the ith node, in addition to the vertical deflection w_i, the rotations θ_{x_i} (see footnote) and θ_{yi} of the normal to the middle surfaces

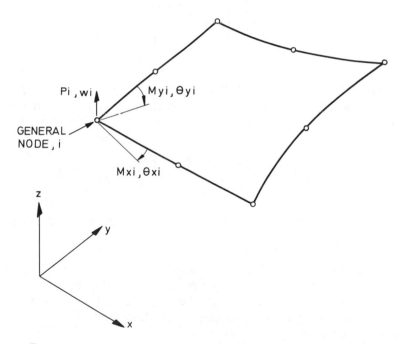

FIG. 1.6. Generalised forces and displacements for plate bending element.

Logically, this should be written θ_{x_i}. However, for convenience, use of a double subscript notation is avoided where possible throughout the text.

in the xz and yz planes respectively are chosen as displacement variables. The nodal forces must then be correspondingly chosen and are termed *generalised forces*. The basic requirement for structural applications is that the product of the generalised displacement and the corresponding generalised force must constitute a work term.

The generalised displacements and forces for each element can then be written in a partitioned form as [1–4]

$$\delta^e = \begin{bmatrix} \delta_1 \\ \delta_2 \\ . \\ . \\ . \\ \delta_n \end{bmatrix} \qquad F^e = \begin{bmatrix} F_1 \\ F_2 \\ . \\ . \\ . \\ F_n \end{bmatrix}, \qquad (1.10)$$

where it is assumed that an element has n nodal points. In plate bending applications, for example, typical terms in (1.10) are

$$\delta_i = \begin{bmatrix} w_i \\ \theta_{xi} \\ \theta_{yi} \end{bmatrix} \qquad F_i = \begin{bmatrix} P_i \\ M_{xi} \\ M_{yi} \end{bmatrix}, \qquad (1.11)$$

where P_i is the nodal load applied normal to the plate and M_{xi} and M_{yi} are nodal couples applied in the xz and yz planes respectively as shown in Fig. 1.6.

For two-dimensional elasticity the generalised forces and displacements are simply

$$\delta_i = \begin{bmatrix} u_i \\ v_i \end{bmatrix} \qquad F_i = \begin{bmatrix} P_{xi} \\ P_{yi} \end{bmatrix}, \qquad (1.12)$$

in which u_i, v_i and P_{xi}, P_{yi} are the Cartesian components of nodal displacements and nodal forces respectively.

Having thus established the nodal displacements, the displacements at any point inside the element are expressed in terms of these through a set of algebraic functions, N_i, which are generally functions of the spatial co-ordinates. These functions are termed the shape functions and are dealt with fully in Chapters 4 and 5. Thus the displacement at a point within the element can be expressed as

$$\delta = \begin{bmatrix} u \\ v \end{bmatrix} = N\delta^e = \sum_{i=1}^{n} N_i \delta_i, \qquad (1.13)$$

where

$$\mathbf{N} = [\mathbf{N}_1, \mathbf{N}_2, \ldots, \mathbf{N}_n] \tag{1.14}$$

and $\mathbf{N}_i = N_i\mathbf{I}$ where $\mathbf{I}$ is an $r \times r$ identity matrix with r being the number of variables per node. Any function N_i must clearly supply a unit value when the coordinate values of node i are substituted and zero when the coordinates of any other node are inserted.

Before determining the relationship between generalised forces and displacements, it is first necessary to establish the strains in terms of displacements and then to examine the interdependence of stresses and strains.

1.5.2 Strain–Displacement relationship

The strains within the element are readily expressed in terms of the displacements or their derivatives. For example, in plane stress situations the strains are defined as

$$\mathbf{\varepsilon} = \begin{bmatrix} \varepsilon_x \\ \varepsilon_y \\ \gamma_{xy} \end{bmatrix} = \begin{bmatrix} \partial u/\partial x \\ \partial v/\partial y \\ \dfrac{\partial u}{\partial y} + \dfrac{\partial v}{\partial x} \end{bmatrix}. \tag{1.15}$$

Substitution for the displacements from (1.13) results in

$$\mathbf{\varepsilon} = \mathbf{B}\mathbf{\delta}^e = \sum_{i=1}^{n} \mathbf{B}_i\mathbf{\delta}_i, \tag{1.16}$$

where

$$\mathbf{B}_i = \begin{bmatrix} \dfrac{\partial N_i}{\partial x} & 0 \\ 0 & \dfrac{\partial N_i}{\partial y} \\ \dfrac{\partial N_i}{\partial y} & \dfrac{\partial N_i}{\partial x} \end{bmatrix}. \tag{1.17}$$

Thus the element strains are expressed directly in terms of the nodal displacements, by means of the **B** matrix which is appropriately termed the strain matrix.

1.5.3 Constitutive law

The relationship between stresses and strains again depends on the application envisaged. For plane stress situations it is easily verified, [8], that the strains can be expressed in terms of the stress components as

$$\varepsilon_x - \varepsilon_x^0 = \frac{1}{E}\sigma_x - \frac{v}{E}\sigma_y$$

$$\varepsilon_y - \varepsilon_y^0 = \frac{1}{E}\sigma_y - \frac{v}{E}\sigma_x$$

$$\gamma_{xy} - \gamma_{xy}^0 = \frac{2(1 + v)}{E}\tau_{xy} \tag{1.18}$$

where σ_x, σ_y, τ_{xy} are the stress components, ε_x^0, ε_y^0, γ_{xy}^0 are initial strains existing in the solid and E and v are the elastic modulus and Poisson's ratio respectively. The initial strains may for example be interpreted as thermal strains or dislocations. When (1.18) is solved for the stresses and any initial stress distribution $\boldsymbol{\sigma}^0$ present in the body before loading is added, the resulting stresses can be written

$$\boldsymbol{\sigma} = \mathbf{D}(\boldsymbol{\varepsilon} - \boldsymbol{\varepsilon}^0) + \boldsymbol{\sigma}^0, \tag{1.19}$$

where

$$\mathbf{D} = \frac{E}{1 - v^2}\begin{bmatrix} 1 & v & 0 \\ v & 1 & 0 \\ 0 & 0 & \dfrac{1 - v}{2} \end{bmatrix}, \tag{1.20}$$

and

$$[\boldsymbol{\sigma}^0]^{\mathrm{T}} = [\sigma_x^0, \sigma_y^0, \tau_{xy}^0]. \tag{1.21}$$

Thus the stresses are completely defined in terms of strains in (1.19) through the **D** matrix.

1.5.4 Equilibrium equations—virtual work

Consider the equilibrium of an element which, in addition to forces applied at the nodes, is also subjected to distributed body forces. In practice these may be, for example, gravity loading or centrifugal effects. These distributed forces may be represented in plane stress situations by

$$\mathbf{p} = \begin{bmatrix} X \\ Y \end{bmatrix}, \tag{1.22}$$

where X and Y are the cartesian body force components per unit volume of material.

The governing equations of an element have already been derived in Section 1.2 by the minimisation of a functional, which can be interpreted as the total potential energy for structural applications. An alternative approach is offered by a straightforward application of the theorem of virtual work and the equations will be derived again using this principle, with initial stresses and strains now being included. Consider a single element acted upon by nodal loads $\mathbf{F}^e$ and body forces $\mathbf{p}$ which result in an equilibrating stress field $\boldsymbol{\sigma}$. Suppose that this element is subjected to an arbitrary virtual nodal displacement pattern $\boldsymbol{\delta}_*^e$ which results in compatible internal displacement and strain distributions of $\boldsymbol{\delta}_*$ and $\boldsymbol{\varepsilon}_*$ respectively. Then the principle of virtual work requires that

$$[\boldsymbol{\delta}_*^e]^T \mathbf{F}^e + \int_{V_e} [\boldsymbol{\delta}_*]^T \mathbf{p} \, dV = \int_{V_e} [\boldsymbol{\varepsilon}_*]^T \boldsymbol{\sigma} \, dV, \tag{1.23}$$

where integration is over the element volume. Use of (1.13) and (1.16) results in

$$[\boldsymbol{\delta}_*^e]^T \left\{ \mathbf{F}^e + \int_{V_e} [\mathbf{N}]^T \mathbf{p} \, dV \right\} = [\boldsymbol{\delta}_*^e]^T \int_{V_e} [\mathbf{B}]^T \boldsymbol{\sigma} \, dV. \tag{1.24}$$

Since the virtual nodal displacement system is arbitrary, the above expression must hold for all values of $\boldsymbol{\delta}_*^e$. Hence

$$\mathbf{F}^e + \int_{V_e} [\mathbf{N}]^T \mathbf{p} \, dV = \int_{V_e} [\mathbf{B}]^T \boldsymbol{\sigma} \, dV. \tag{1.25}$$

Substituting for $\boldsymbol{\sigma}$ from (1.19)

$$\mathbf{F}^e + \int_{V_e} [\mathbf{N}]^T \mathbf{p} \, dV = \left\{ \int_{V_e} [\mathbf{B}]^T \mathbf{D} \mathbf{B} \, dV \right\} \boldsymbol{\delta}^e$$

$$- \int_{V_e} [\mathbf{B}]^T \mathbf{D} \boldsymbol{\varepsilon}^0 \, dV + \int_{V_e} [\mathbf{B}]^T \boldsymbol{\sigma}^0 \, dV, \tag{1.26}$$

or

$$\mathbf{F}^e + \mathbf{F}_p^e + \mathbf{F}_{\varepsilon^0}^e + \mathbf{F}_{\sigma^0}^e = \mathbf{K}^e \boldsymbol{\delta}^e, \tag{1.27}$$

where

$$\mathbf{K}^e = \int_{V_e} [\mathbf{B}]^T \mathbf{D} \mathbf{B} \, dV, \tag{1.28}$$

$$\mathbf{F}_p^e = \int_{V_e} [\mathbf{N}]^T \mathbf{p} \, dV, \tag{1.29}$$

$$\mathbf{F}_{\varepsilon^0}^e = \int_{V_e} [\mathbf{B}]^T \mathbf{D} \varepsilon^0 \, dV, \tag{1.30}$$

$$\mathbf{F}_{\sigma^0}^e = - \int_{V_e} [\mathbf{B}]^T \sigma^0 \, dV. \tag{1.31}$$

It should be emphasised that (1.23)–(1.27) are only true if a single element is considered. For more than one element, all terms in these equations should be summed over all the elements present in the structure. Expressions (1.28)–(1.31) define respectively the element stiffness matrix, the equivalent nodal forces for body force, initial strain and initial stress loadings. It is noted that the element stiffness matrix (1.28) is identical to that obtained by the functional approach in (1.9) and that the nodal force expression for body forces (1.29) agrees with the corresponding term in (1.8).

As indicated in Section 1.2 the stiffness equations (1.27) are assembled for the complete structure and solved for the nodal displacements. On use of (1.16) and (1.19), the element stresses are obtained from the relation

$$\sigma = \mathbf{D}(\mathbf{B}\delta^e - \varepsilon^0) + \sigma^0. \tag{1.32}$$

The assembly of element equations is discussed in detail in Chapter 2.

1.5.5 Distributed loading on boundaries

Suppose that an element face is subjected to traction forces

$$\mathbf{t} = \begin{bmatrix} t_x \\ t_y \end{bmatrix}, \tag{1.33}$$

where t_x and t_y are the force components/unit surface area in the x and y directions respectively. An additional term of

$$\int_{S_e} [\delta_*]^T \mathbf{t} \, dS, \tag{1.34}$$

now appears on the left hand side of equation (1.23). The integration in (1.34) is taken over the loaded element surface. This results in an additional equivalent nodal force term $\mathbf{F}_t^e$ occurring in the left hand side of equation (1.27), where

$$\mathbf{F}_t^e = \int_{S_e} [\mathbf{N}]^T \mathbf{t} \, dS. \tag{1.35}$$

Thus distributed surface loadings are treated as equivalent nodal forces. Again the contribution of each element should be summed, if more than one element is present.

Prescribed displacement boundary conditions are accommodated by direct introduction of the specified values at nodal points after assembly of the final equations. The procedure employed for this is discussed in detail in Chapters 2 and 8.

1.6 Finite element program structure

As mentioned in Section 1.3 a modular approach is adopted for the programs presented in this text, with the various main finite element operations being performed by separate subroutines. Fig. 1.7 shows the organisation of all programs presented, particularly the sequence in which the subroutines are accessed. The basic finite element steps are performed by *primary subroutines* which rely on *auxiliary subroutines* to carry out secondary operations. An auxiliary subroutine may be required by more than one primary subroutine as shown in Fig. 1.7. The order of calling of the primary subroutines is controlled by a main or master segment. The function of each subroutine is described in the remainder of this section. It can be seen in Fig. 1.7 that a suffix *B*, *PS* of *PB* appears in some subroutine names: These refer to beam, plane stress/strain and plate bending applications respectively. Similarly a suffix 1 or 2 is employed to differentiate between one-dimensional geometric situations (beam) and two-dimensional problems (plant stress/strain and plate bending).

The construction of finite element programs employing the displacement approach falls naturally into three phases.

Phase 1. Semi-theoretical aspects such as input and output
Whilst being possibly the least technologically demanding aspect of a finite element analysis, from a practical engineering viewpoint, input/output is arguably the most important. The program subroutine controlling the input is considered in detail in Chapter 3. The data input subroutine is named INPUT (See Fig. 1.7) and is common to all applications presented.

A separate subroutine is not employed to output the results. Instead the results are output as soon as they are obtained. The displacements are output in the equation solution subroutine FRONT and the stress components are output from the stress evaluation subroutines which are titled STREB, STREPS, STREPB for the beam, plane stress/strain and plate bending cases respectively.

Phase 2. Stiffness and stress matrices and applied load vector generation
It is at this stage that the main use is made of the basic expressions of finite element theory. For structural analysis by the displacement approach, the process clearly follows the steps taken in the matrix methods of structural frame analysis.

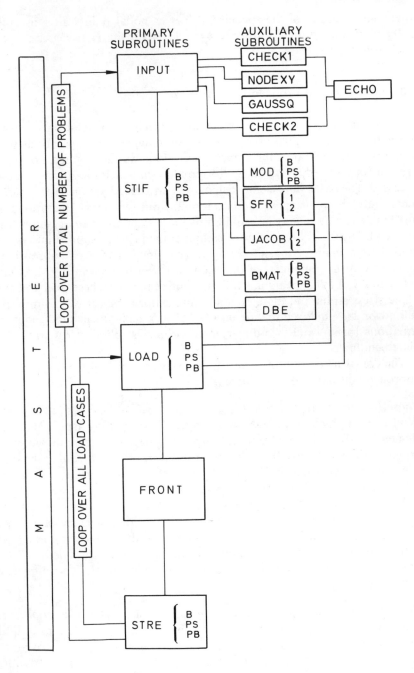

Fig. 1.7. Program organisation.

The stiffness and stress matrices for the beam are calculated in Chapter 4, for plane stress/strain conditions in Chapter 5 and for plate bending situations in Chapter 6. The subroutines which perform this task for the beam, plane and plate applications are named STIFB, STIFPS and STIFPB respectively. After subsequent solution for the nodal displacements, the element stress matrices are employed in the evaluation of the stress components or stress resultants. This task is performed for the beam, plane and plate elements respectively by subroutines STREB, STREPS and STREPB.

The displacement method of finite element analysis relies on all structural loading being interpreted as equivalent nodal forces as described in Sections 1.5.4 and 1.5.5. For example, gravity loading or pressures applied to element faces must be converted to equivalent discrete nodal forces. The subroutines which accomplish this, as well as accepting the loading data, for the beam, plane and plate elements respectively, are LOADB, LOADPS and LOADPB. These are described in detail in Chapter 4 for the beam element and Chapter 7 for both the plane stress/strain and plate bending elements.

Phase 3. Solution of the stiffness equations.
The time spent on the solution of the stiffness equations of the structure represents a large percentage of the total computation time. Therefore the method of solving these equations is critical to an efficient solution. Whilst there are several ways in which the operations outlined in Phases 1 and 2 can be performed, the optimal approach will only produce marginal savings in both computer core storage and solution costs. However, the equation solution scheme adopted can affect both factors considerably. In the preparation of a text such as this a dilemma immediately arises, since the simplest algorithms are generally inefficient with regard to either storage requirements or computational effort or both. In an attempt to provide programs which are of benefit to the elementary student and also to interest the more sophisticated user, a relatively elementary version of a *frontal process equation solver* is presented. In its present form the program is suitable for teaching or research purposes. Subroutine FRONT, described in Chapter 8, is the equation solution subroutine whose function is to assemble the element stiffness equations and solve for the unknown displacements and reactions using the frontal elimination technique.

As mentioned previously the function of the auxiliary subroutines is to carry out computations required by one or more of the primary subroutines and the task of each one is described below. The location where they are developed in the text is also indicated.

Subroutine NODEXY (chapter 3)
Calculates the coordinates of midside nodes which lie on a straight line

connecting two adjacent corner nodes. For example, if in Fig. 1.5 midside node q is co-linear with corner nodes p and r, specification of the coordinates of node q can be avoided, thus reducing the total amount of input data required. This subroutine checks to see if the coordinates of a midside node have been specified and if not, they are determined by linear interpolation.

Subroutine GAUSSQ (Chapter 3)
Sets up the sampling point positions and weighting factors for numerical integration.

Subroutines MODB, MODPS, MODPB (Chapters 4, 5 and 6)
These subroutines set up the **D** matrix (defined in (1.19)) for beam, plane stress/strain and plate bending applications respectively.

Subroutines SFR1, SFR2 (Chapters 4 and 5)
The shape functions and their local derivatives are computed for one and two-dimensional situations respectively by these routines.

Subroutines JACOB1, JACOB2 (Chapters 4 and 5)
These compute the Jacobian matrix and its inverse and the Cartesian derivatives of the shape functions for one and two-dimensional situations respectively.

Subroutines BMATB, BMATPS, BMATPB (Chapters 4, 5 and 6)
These subroutines compute the strain matrix, **B**, (defined in (1.16)) for beam, plane stress/strain and plate bending applications respectively.

Subroutine DBE (Chapter 4)
Computes the matrix product **DB** required for stress determination.

Subroutine CHECK1 (Chapter 9)
As soon as the control parameters (e.g. number of degrees of freedom per node, etc.) have been read in subroutine INPUT, subroutine CHECK1 is called to scrutinise this section of data. If any errors are detected, diagnostic messages are printed and the remainder of the data echoed before the job is aborted.

Subroutine CHECK2 (Chapter 9)
As soon as the remainder of the data has been assimilated by subroutine INPUT it is checked by subroutine CHECK2. Again, any errors are signalled by diagnostic messages and any remaining data echoed before the job is aborted.

Subroutine ECHO (Chapter 9)

This subroutine merely reads and echoes by lineprinter the remaining data after at least one error has been detected by subroutines CHECK1 or CHECK2.

A User's Manual for data preparation is provided in Appendix I.

1.7 Variable nomenclature

In the programs presented in the remainder of this text an attempt has been made to name variables in a logical manner. By choosing descriptive names, the use of many of the variables becomes self-apparent, thus assisting the reader in the task of program assimilation. All variable names are chosen to be 5 characters in length; this occasionally causes a little difficulty in abbreviation but has an advantage with regard to neatness of program presentation. For example, the following names will be employed.

NMATS	The Number of different MATerialS
PROPS(...)	The array of material PROPertieS
NEVAB	The Number of Element VAriaBles
NNODE	The Number of NODes per Element
NDOFN	The Number of Degrees Of Freedom per Node

Furthermore a "common root" principle will be adopted; where a single basic variable name is employed with different prefixes depending on its usage in the program. In particular:

(i)   Prefix I, J or L will be used to indicate a DO loop variable
(ii) Prefix K will indicate a counter
(iii) Prefix M will indicate a maximum value
(iv) Prefix N will indicate a given number

For example IPOIN, NPOIN, MPOIN will indicate respectively a particular nodal point, the number of nodal points in the problem and the maximum permissible number of nodal points in the program.

Similarly, any DO loop will be of the general form

```
        KEVAB=0
        DO 1 INODE=1,NNODE
        DO 1 IDOFN=1,NDOFN
      1 KEVAB=KEVAB+1
```

which indicates that the outer and inner DO loop indices range respectively over the number of nodes per element and the number of degrees of freedom per node. The prefix K is employed in KEVAB to indicate a counter over the number of element variables, NEVAB.

A dictionary of the variable names employed throughout the text is given in Appendix II for ease of reference.

1.8 Program description scheme

All subroutines which are described in this text are presented in a standard form. An input/output diagram showing the main variables to be transferred to and from the subroutine is provided. The mode (e.g. common statement, subroutine argument) by which the transfer of information is achieved is also indicated. A FORTRAN listing is then presented with detailed notes on each group of statements. Comment cards have also been used to assist in the understanding of the program. Apart from the demonstration subroutines given in Chapter 2, description of the common blocks used in the subroutines has been delayed until Chapter 10. In that chapter, which deals with the assembly of the subroutines to produce the three main programs, the common blocks are described in detail.

At the beginning of Chapters in which subroutines employed in one or more of the three programs are presented for the first time, a diagram is given to remind the reader of the program organisation. These new subroutines are indicated by shading.

References

1. Zienkiewicz, O. C., "The Finite Element Method in Engineering Science". McGraw-Hill, New York, 1971.
2. Desai, C. S., and Abel, J. F., "An Introduction to the Finite Element Method". Van Nostrand Reinhold, New York, 1972.
3. Gallagher, R. H., "Finite Element Analysis—Fundamental". Prentice Hall, New Jersey, 1975.
4. Norrie, D. H., and de Vries, G., "The Finite Element Method—Fundamentals and Applications". Academic Press, London, 1973.
5. Anderson, R. G., Irons, B. M. and Zienkiewicz, O. C., Vibration and stability of plates using finite elements. *Int. J. Solids and Struct.*, **4**, 1031–55, 1968.
6. Taylor, C., and Davis, J. M., *In* "Tidal Propagation and Dispersion in Estuaries". Finite Elements in Flow Problems, Ch. 15, J. T. Oden and R. H. Gallagher, C. Taylor and O. C. Zienkiewicz (eds.), Wiley, London, 1975.
7. Zienkiewicz, O. C., Owen, D. R. J., Phillips, D. V., and Nayak, G. C., Finite element methods in the analysis of reactor vessels. *Nucl. Eng. Des.* **20**, 507–541, 1972.
8. Timoshenko, S., and Goodier, J. N., "Theory of Elasticity". McGraw-Hill, New York, 1951.

2

Assembly of Elements and Solution of Stiffness Equations

2.1 Introduction

In the finite element method individual element stiffness matrices and applied load vectors are calculated separately and then assembled into the overall structural stiffness matrix and load vector respectively. This chapter introduces the basic ideas of element assembly and then discusses some elementary methods of solving the equations thus generated. Some of the special features of these equations are then described to prepare the reader for the solution scheme presented in Chapter 8. Throughout the present chapter, small subroutines are presented for the assembly process and the elementary equation solving procedures. It should be emphasised that these subroutines will *not* be used in the main finite element program.

2.2 Element assembly for a one-dimensional axial load member

In order to introduce the element assembly process we consider the simple example of a one-dimensional axial load member divided into 3 elements as shown in Fig. 2.1. In this example, we use quadratic isoparametric elements with three nodes—one at each end of and one halfway along the element. At each node i of the element there is an axial displacement degree of freedom u_i.

We assume that the element stiffness matrices and the element load vectors

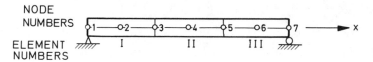

FIG. 2.1. Assemblage of 1-D axial elements.

are known and may be written as follows

$$\mathbf{K}^{I} = \begin{bmatrix} a_{11} & a_{12} & a_{13} \\ a_{12} & a_{22} & a_{23} \\ a_{13} & a_{23} & a_{33} \end{bmatrix}, \qquad \mathbf{K}^{II} = \begin{bmatrix} b_{11} & b_{12} & b_{13} \\ b_{12} & b_{22} & b_{23} \\ b_{13} & b_{23} & b_{33} \end{bmatrix},$$

$$\mathbf{K}^{III} = \begin{bmatrix} c_{11} & c_{12} & c_{13} \\ c_{12} & c_{22} & c_{23} \\ c_{13} & c_{23} & c_{33} \end{bmatrix},$$

$$\mathbf{F}^{I} = \begin{bmatrix} A_1 \\ A_2 \\ A_3 \end{bmatrix}, \qquad \mathbf{F}^{II} = \begin{bmatrix} B_1 \\ B_2 \\ B_3 \end{bmatrix}, \qquad \mathbf{F}^{III} = \begin{bmatrix} C_1 \\ C_2 \\ C_3 \end{bmatrix}. \qquad (2.1)$$

These are particular cases of the expressions given in (1.28)–(1.31). In passing we note the symmetry of the above matrices. The vectors of the unknown nodal displacements for the elements are

$$\boldsymbol{\delta}^{I} = \begin{bmatrix} u_1 \\ u_2 \\ u_3 \end{bmatrix}, \qquad \boldsymbol{\delta}^{II} = \begin{bmatrix} u_3 \\ u_4 \\ u_5 \end{bmatrix}, \qquad \boldsymbol{\delta}^{III} = \begin{bmatrix} u_5 \\ u_6 \\ u_7 \end{bmatrix}. \qquad (2.2)$$

We now use the Theorem of Minimum Total Potential Energy to derive the stiffness equations for this problem. The total potential energy for each element may be calculated separately. For example, the total potential energy of element I can be expressed as

$$\pi^{I} = \tfrac{1}{2}[\boldsymbol{\delta}^{I}]^{T}\mathbf{K}^{I}\boldsymbol{\delta}^{I} - [\boldsymbol{\delta}^{I}]^{T}\mathbf{F}^{I}$$

$$= \tfrac{1}{2}[u_1(a_{11}u_1 + a_{12}u_2 + a_{13}u_3) + u_2(a_{12}u_1 + a_{22}u_2 + a_{23}u_3)$$

$$+ u_3(a_{13}u_1 + a_{23}u_2 + a_{33}u_3)] - [A_1u_1 + A_2u_2 + A_3u_3]. \qquad (2.3)$$

The total potential energy for the assemblage is given by the sum of the individual element potentials

$$\pi = \pi^{\mathrm{I}} + \pi^{\mathrm{II}} + \pi^{\mathrm{III}}. \qquad (2.4)$$

Using the principle of minimum potential energy, we obtain

$$
\left.
\begin{aligned}
\frac{\partial \pi}{\partial u_1} &= a_{11}u_1 + a_{12}u_2 + a_{13}u_3 - A_1 = 0 \\[2mm]
\frac{\partial \pi}{\partial u_2} &= a_{12}u_1 + a_{22}u_2 + a_{33}u_3 - A_2 = 0 \\[2mm]
\frac{\partial \pi}{\partial u_3} &= a_{13}u_1 + a_{23}u_2 + (a_{33}+b_{11})u_3 + b_{12}u_4 + b_{13}u_5 - (A_3 + B_1) = 0 \\[2mm]
\frac{\partial \pi}{\partial u_4} &= b_{12}u_3 + b_{22}u_4 + b_{23}u_5 - B_2 = 0 \\[2mm]
\frac{\partial \pi}{\partial u_5} &= b_{13}u_3 + b_{23}u_4 + (b_{33}+c_{11})u_5 + c_{12}u_6 + c_{13}u_7 - (C_1 + B_3) = 0 \\[2mm]
\frac{\partial \pi}{\partial u_6} &= c_{12}u_5 + c_{22}u_6 + c_{23}u_7 - C_2 = 0 \\[2mm]
\frac{\partial \pi}{\partial u_7} &= c_{13}u_5 + c_{23}u_6 + c_{33}u_7 - C_3 = 0.
\end{aligned}
\right\} \qquad (2.5)
$$

These equilibrium equations for the asemblage can be expressed in matrix form as

$$
\begin{array}{c}
\begin{array}{ccccccc} 1 & 2 & 3 & \quad 4 & 5 & \quad 6 & 7 \end{array} \\
\begin{array}{c}1\\2\\3\\4\\5\\6\\7\end{array}
\left[
\begin{array}{ccccccc}
a_{11} & a_{12} & a_{13} & & & & \\
a_{12} & a_{22} & a_{23} & & & & \\
a_{13} & a_{23} & a_{33}+b_{11} & b_{12} & b_{13} & & \\
 & & b_{12} & b_{22} & b_{23} & & \\
 & & b_{13} & b_{23} & b_{33}+c_{11} & c_{12} & c_{13} \\
 & & & & c_{12} & c_{22} & c_{23} \\
 & & & & c_{13} & c_{23} & c_{33}
\end{array}
\right]
\end{array}
\begin{bmatrix} u_1 \\ u_2 \\ u_3 \\ u_4 \\ u_5 \\ u_6 \\ u_7 \end{bmatrix}
=
\begin{bmatrix} A_1 \\ A_2 \\ A_3+B_1 \\ B_2 \\ C_1+B_3 \\ C_2 \\ C_3 \end{bmatrix}
$$

or

$$\mathbf{K}\boldsymbol{\delta} = \mathbf{F}. \qquad (2.6)$$

The assembly procedure can readily be appreciated by comparing the equations for the assemblage, (2.6) with those for the individual elements, (2.1). Clearly, the individual element stiffnesses and load vectors can be added directly to the overall matrix **K** and load vector **F** in positions appropriate to the node numbers of the elements.

2.3 Element assembly for a 2D Isoparametric element

Our next example is slightly more complex. We assume for ease of explanation that only one degree of freedom exists at each node. (This is common in non-structural applications, e.g. temperature in a heat flow problem.)

We shall consider the assemblage of 2D elements in Fig. 2.2. Each element is an 8-noded isoparametric element, with four corner nodes and four midside nodes. The first element is shown in Fig. 2.3. We can specify its stiffness matrix by writing only the upper triangle since this matrix is symmetric:

$$(2.7)$$

global	local	11	7	3	2	1	6	9	10	global
		1	2	3	4	5	6	7	8	local
11	1	$K_{1,1}$	$K_{1,2}$	$K_{1,3}$	$K_{1,4}$	$K_{1,5}$	$K_{1,6}$	$K_{1,7}$	$K_{1,8}$	Q_1
7	2		$K_{2,2}$	$K_{2,3}$	$K_{2,4}$	$K_{2,5}$	$K_{2,6}$	$K_{2,7}$	$K_{2,8}$	Q_2
3	3			$K_{3,3}$	$K_{3,4}$	$K_{3,5}$	$K_{3,6}$	$K_{3,7}$	$K_{3,8}$	Q_3
2	4				$K_{4,4}$	$K_{4,5}$	$K_{4,6}$	$K_{4,7}$	$K_{4,8}$	Q_4
1	5					$K_{5,5}$	$K_{5,6}$	$K_{5,7}$	$K_{5,8}$	Q_5
6	6						$K_{6,6}$	$K_{6,7}$	$K_{6,8}$	Q_6
9	7							$K_{7,7}$	$K_{7,8}$	Q_7
10	8								$K_{8,8}$	Q_8

stiffness for element I   load vector for element I

If the components of the element stiffness matrix are now inserted into the given rows and columns of the overall stiffness matrix we obtain the following result

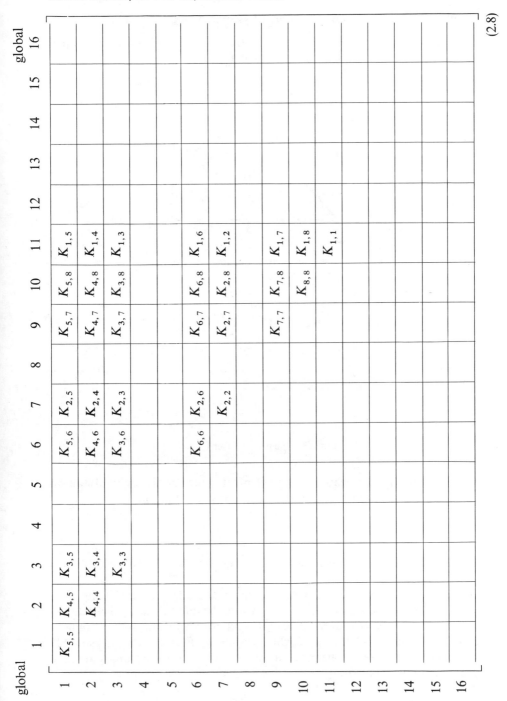

(2.8)

Similarly, the transpose of the overall assembled load vector is

1   2   3   4   5   6   7   8   9   10   11   12 13 14 15 16 global
$[Q_5 \quad Q_4 \quad Q_3 \qquad\qquad Q_6 \quad Q_2 \qquad Q_7 \quad Q_8 \quad Q_1 \qquad\qquad\qquad\qquad]$

$$(2.9)$$

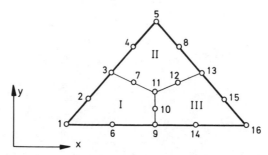

FIG. 2.2.   Assemblage of 2-D isoparametric elements.

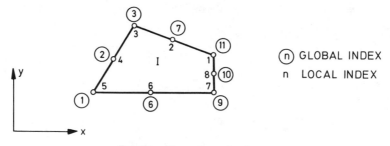

FIG. 2.3.   Element number I.

This process may be repeated for each element in turn until the global stiffness matrix and load vector are complete, resulting in a set of linear equations to be solved.

2.4 Element assembly as a FORTRAN operation—subroutine ASSEMB

Object: As we shall see in a Chapter 8, we never actually assemble the complete global stiffness matrix, as this would be unnecessarily wasteful in computer storage. However, as an introduction, we shall now examine a demonstration assembly routine which restates in FORTRAN the assembly process demonstrated in the previous section. The variables employed in the program are listed below.

Dictionary of variable names (with dimensions)

ASLOD(MSVAB)	ASsembled LOaD vector
ASTIF(MSVAB,MSVAB)	Assembled global STIFfness matrix
RLOAD(MEVAB)	Element load vector
ESTIF(MEVAB,MEVAB)	Element STIFfness matrix
IELEM,NELEM,MELEM	Index, Number, Maximum of ELEMents
IFILE	Input FILE
IDOFN,JDOFN,NODFN	Index, Index, Number of Degrees Of Freedom per Node
INODE,JNODE,NNODE,MNODE	Index, Index, Number, Maximum of NODes per Element
ISVAB,JSVAB,MSVAB,NSVAB	Index, Index, Maximum, Number of Global Structural VAriaBles
JFILE	Output file
LNODS(MELEM,MNODE)	ELement NODe numberS listed for each element
NODEI	NODE I
NODEJ	NODE J
NCOLS	Number of the COLumn in the global Structural stiffness matrix
NROWS	Number of the ROW in the global Structural stiffness matrix and load vector
NCOLE	Number of the COLumn in the Element stiffness matrix
NROWE	Number of the ROW in the Element stiffness matrix and load vector
MEVAB	Maximum of Element VAriaBles

We indicate what must be input to the subroutine and what is output from the subroutine by an *Input/Output* diagram

```
              INPUT                                OUTPUT
           ┌ IFILE                              ┌ ASTIF(MSVAB,MSVAB) ┐
           │ MSVAB          →│ ASSEMB │→        │ ASLOD(MSVAB)       │ A
           │ MEVAB                              └                    ┘
    A      │ MELEM
           │ MNODE
           └ LNODS(MELEM,MNODE)

    F(IFILE) ┌ ESTIF(MEVAB,MEVAB)
             └ RLOAD(MEVAB)
```

$$C(A1) \quad \begin{bmatrix} NSVAB \\ NELEM \\ NNODE \\ NDOFN \end{bmatrix}$$

Note that A means information is passed through the argument of the subroutine, F means information is transmitted by use of a file and C means information is transmitted by use of common blocks. The symbols A, F and C are used continuously throughout the book.

Annotated FORTRAN listing

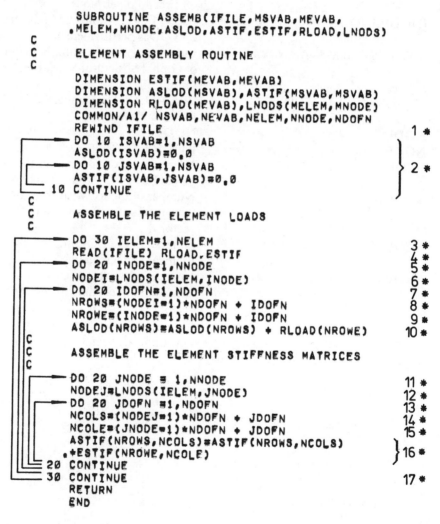

```
      SUBROUTINE ASSEMB(IFILE,MSVAB,MEVAB,
     .MELEM,MNODE,ASLOD,ASTIF,ESTIF,RLOAD,LNODS)
C
C     ELEMENT ASSEMBLY ROUTINE
C
      DIMENSION ESTIF(MEVAB,MEVAB)
      DIMENSION ASLOD(MSVAB),ASTIF(MSVAB,MSVAB)
      DIMENSION RLOAD(MEVAB),LNODS(MELEM,MNODE)
      COMMON/A1/ NSVAB,NEVAB,NELEM,NNODE,NDOFN
      REWIND IFILE                                       1 *
      DO 10 ISVAB=1,NSVAB
      ASLOD(ISVAB)=0.0
      DO 10 JSVAB=1,NSVAB                                2 *
      ASTIF(ISVAB,JSVAB)=0.0
   10 CONTINUE
C
C     ASSEMBLE THE ELEMENT LOADS
C
      DO 30 IELEM=1,NELEM                                3 *
      READ(IFILE) RLOAD,ESTIF                            4 *
      DO 20 INODE=1,NNODE                                5 *
      NODEI=LNODS(IELEM,INODE)                           6 *
      DO 20 IDOFN=1,NDOFN                                7 *
      NROWS=(NODEI-1)*NDOFN + IDOFN                      8 *
      NROWE=(INODE-1)*NDOFN + IDOFN                      9 *
      ASLOD(NROWS)=ASLOD(NROWS) + RLOAD(NROWE)          10 *
C
C     ASSEMBLE THE ELEMENT STIFFNESS MATRICES
C
      DO 20 JNODE = 1,NNODE                             11 *
      NODEJ=LNODS(IELEM,JNODE)                          12 *
      DO 20 JDOFN =1,NDOFN                              13 *
      NCOLS=(NODEJ-1)*NDOFN + JDOFN                     14 *
      NCOLE=(JNODE-1)*NDOFN + JDOFN                     15 *
      ASTIF(NROWS,NCOLS)=ASTIF(NROWS,NCOLS)
     .+ESTIF(NROWE,NCOLE)                               16 *
   20 CONTINUE
   30 CONTINUE                                          17 *
      RETURN
      END
```

1*   Rewind file ready for reading and writing.
2*   Zero ASLOD and ASTIF.
3*   Loop for each element.
4*   Read ESTIF and RLOAD for the current element.
5*   Loop for each node 'INODE' of current element.
6*   From LNODS array identify node number of current node 'INODE'.
7*   Loop for each degree of freedom of the current node 'INODE'.
8*   Establish row position in global stiffness matrix and load vector.
9*   Establish row position in element stiffness matrix and load vector.
10* Add contribution to global load vector from element load vector.
11* Loop for each node 'JNODE' of current element.
12* From LNODS array identify node number of current node 'JNODE'.
13* Loop for each degree of freedom of the current node 'JNODE'.
14* Establish column position in global stiffness matrix.
15* Establish column position in element stiffness matrix.
16* Add contribution to global stiffness matrix from element stiffness matrix.
17* End element loop.

General notes
For the problem described in Section 2.2, the main variables have the following values

$$NSVAB = 7, NNODE = 3, NELEM = 3, NDOFN = 1$$

$$LNODS = \begin{bmatrix} 1 & 2 & 3 \\ 3 & 4 & 5 \\ 5 & 6 & 7 \end{bmatrix} \begin{matrix} —\text{1st element} \\ —\text{2nd element} \\ —\text{3rd element} \end{matrix}$$

Similarly, for the problem described in Section 2.3

$$NSVAB = 16, NNODE = 8, NELEM = 3, NDOFN = 1,$$

$$LNODS = \begin{bmatrix} 11 & 7 & 3 & 2 & 1 & 6 & 9 & 10 \\ 11 & 12 & 13 & 8 & 5 & 4 & 3 & 7 \\ 9 & 14 & 16 & 15 & 13 & 12 & 11 & 10 \end{bmatrix} \begin{matrix} —\text{1st element} \\ —\text{2nd element} \\ —\text{3rd element} \end{matrix}$$

Here, for ease of explanation, we have only dealt with 1 degree of freedom per node. The reader is advised to repeat the problems in Sections 2.2 and 2.3 with 2 degrees of freedom per node using the FORTRAN subroutine as a guide.

2.5 The solution of equations

In this section we shall introduce some elementary concepts of equation solving. Firstly, we shall consider Gauss–Jordan reduction which is the most fundamental method of equation solving. We shall then proceed to Gaussian reduction which will be used in a more efficient form in the main solution routine described later in Chapter 8. We note that at this stage we are taking no advantage of the probable 'banded' nature of the global stiffness matrix. An introduction to this facet of equation solving will be presented in Section 2.6.

2.5.1 Gauss–Jordan reduction

Suppose we have a set of equations with unknowns u_1, u_2 and u_3

$$3u_1 + u_2 + 2u_3 = 11 \tag{2.10a}$$

$$u_1 + u_2 + u_3 = 6 \tag{2.10b}$$

$$2u_1 + u_2 + 2u_3 = 10. \tag{2.10c}$$

To solve these equations (2.10) using Gauss–Jordan reduction, we first eliminate u_1 from all equations except (a). Then we eliminate u_2 from all equations except (b) and finally we eliminate u_3 from all equations except (c).

To eliminate u_1, we subtract $\frac{1}{3}$ of (2.10a) from (2.10b) and $\frac{2}{3}$ of (2.10a) from (2.10c) which results in the following equations

$$3u_1 + u_2 + 2u_3 = 11 \tag{2.11a}$$

$$0u_1 + \tfrac{2}{3}u_2 + \tfrac{1}{3}u_3 = \tfrac{7}{3} \tag{2.11b}$$

$$0u_1 + \tfrac{1}{3}u_2 + \tfrac{2}{3}u_3 = \tfrac{8}{3}. \tag{2.11c}$$

To eliminate u_2, we subtract $\frac{3}{2}$ of (2.11b) from (2.11a) and $\frac{1}{2}$ of (2.11b) from (2.11c) which results in the equations

$$3u_1 + 0u_2 + \tfrac{3}{2}u_3 = \tfrac{15}{2} \tag{2.12a}$$

$$0u_1 + \tfrac{2}{3}u_2 + \tfrac{1}{3}u_3 = \tfrac{7}{3} \tag{2.12b}$$

$$0u_1 + 0u_2 + \tfrac{1}{2}u_3 = \tfrac{3}{2}. \tag{2.12c}$$

Finally, to eliminate u_3, we subtract $\frac{2}{3}$ of (2.12c) from (2.12b) and three times (2.12c) from (2.12a):

$$3u_1 + 0u_2 + 0u_3 = 3 \tag{2.13a}$$

$$0u_1 + \tfrac{2}{3}u_2 + 0u_3 = \tfrac{4}{3} \tag{2.13b}$$

$$0u_1 + 0u_2 + \tfrac{1}{2}u_3 = \tfrac{3}{2} \tag{2.13c}$$

By inspection we see that $u_1 = 1$, $u_2 = 2$ and $u_3 = 3$.

We now summarise in matrix form the above result:

$$\begin{bmatrix} 3 & 1 & 2 \\ 1 & 1 & 1 \\ 2 & 1 & 2 \end{bmatrix} \begin{bmatrix} u_1 \\ u_2 \\ u_3 \end{bmatrix} = \begin{bmatrix} 11 \\ 6 \\ 10 \end{bmatrix}$$

has been transformed into

$$\begin{bmatrix} 3 & 0 & 0 \\ 0 & \frac{2}{3} & 0 \\ 0 & 0 & \frac{1}{2} \end{bmatrix} \begin{bmatrix} u_1 \\ u_2 \\ u_3 \end{bmatrix} = \begin{bmatrix} 3 \\ \frac{4}{3} \\ \frac{3}{2} \end{bmatrix}$$

2.5.2 Subroutine JORDAN

Object: We now examine a Gauss–Jordan reduction routine which restates in FORTRAN the equation solving technique we have just considered. We use the notation adopted in the previous section on assembly.

Dictionary of variable names

ASLOD(MEQNS)	ASembled LOaD vector
ASTIF(MEQNS,MEQNS)	Assembled global STIFfness matrix
IEQNS,NEQNS,MEQNS	Index, Number, Maximum of EQuatioNS
ICOLS	Index COLumn of Structural stiffness matrix
IROWS	Index ROW of Structural stiffness matrix and load vector
FACTR	Jordan reduction FACToR
PIVOT	Diagonal term of variable which is currently being eliminated.

Input/Output diagram

```
         INPUT                                          OUTPUT
       ┌ MEQNS                    →│ JORDAN │→   Modified                  ┐
   A   │ ASTIF(MEQNS,MEQNS)                       ASTIF(MEQNS,MEQNS)  │ A
       └ ASLOD(MEQNS)                             Modified            ┘
                                                  ASLOD(MEQNS)
   C(A1)   [NEQNS
```

Annotated FORTRAN listing

```
        SUBROUTINE JORDAN(MEQNS,ASLOD,ASTIF)
C
C       GAUSS-JORDAN REDUCTION ROUTINE
C
        DIMENSION ASLOD(MEQNS),ASTIF(MEQNS,MEQNS)
        COMMON/A1/ NEQNS,NEVAB,NELEM,NNODE,NDOFN
```

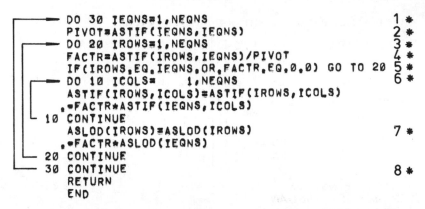

```
      ┌──── DO 30 IEQNS=1,NEQNS                              1 *
      │     PIVOT=ASTIF(IEQNS,IEQNS)                         2 *
      │ ┌── DO 20 IROWS=1,NEQNS                              3 *
      │ │   FACTR=ASTIF(IROWS,IEQNS)/PIVOT                   4 *
      │ │   IF(IROWS.EQ.IEQNS.OR.FACTR.EQ.0.0) GO TO 20      5 *
      │ │ ┌ DO 10 ICOLS=    1,NEQNS                          6 *
      │ │ │ ASTIF(IROWS,ICOLS)=ASTIF(IROWS,ICOLS)
      │ │ │ .=FACTR*ASTIF(IEQNS,ICOLS)
      │ │ └ 10 CONTINUE
      │ │   ASLOD(IROWS)=ASLOD(IROWS)                        7 *
      │ │   .=FACTR*ASLOD(IEQNS)
      │ └── 20 CONTINUE
      └──── 30 CONTINUE                                      8 *
            RETURN
            END
```

1*   Loop for each equation—this equation is associated with variable to be eliminated.

2*   Extract the PIVOT—the leading diagonal term for the variable currently being eliminated.

3*   Loop for each equation 'IROWS'.

4*   Calculate the Gauss–Jordan reduction factor—this is used to make coefficients in column 'IEQNS' zero except for row 'IEQNS'.

5*   If this factor is zero or if 'IROWS' equals 'IEQNS' do not process equation 'IROWS' instead skip to next equation.

6*   Treat each term in equation 'IROWS'.

7*   Treat each R.H.S. (right hand side)

8*   End equation loop.

General notes

For the problem described in Section 2.5.1, the main variables have the following values:

$$NEQNS = 3$$

$$ASLOD = \begin{bmatrix} 11 \\ 6 \\ 10 \end{bmatrix}, \text{modified ASLOD} = \begin{bmatrix} 3 \\ \frac{4}{3} \\ \frac{3}{2} \end{bmatrix}$$

$$ASTIF = \begin{bmatrix} 3 & 1 & 2 \\ 1 & 1 & 1 \\ 2 & 1 & 2 \end{bmatrix}, \text{ modified ASTIF} = \begin{bmatrix} 3 & 0 & 0 \\ 0 & \frac{2}{3} & 0 \\ 0 & 0 & \frac{1}{3} \end{bmatrix}.$$

ASTIF and ASLOD undergo a series of transformations as we eliminate each variable in turn.

The only division in the routine is in 4* by PIVOT which picks up the value taken, at various times, by the three diagonal terms of ASTIF. Evidently, the three values must all be non-zero.

2.5.3 Prescribed values

In virtually every finite element problem many of the variables have prescribed values. Associated with each of these prescribed values is an unknown reaction, R. Let us consider a variant of the previous example in which u_2 has a prescribed value of 4, so that

$$3u_1 + u_2 + 2u_3 = 11 \tag{2.14a}$$

$$u_1 + u_2 + u_3 = 6 + R_2, u_2 = 4 \tag{2.14b}$$

$$2u_1 + u_2 + 2u_3 = 10. \tag{2.14c}$$

We note that (2.14a) and (2.14c) are the same as before but that the R.H.S. of (2.14b) is now $6 + R_2$, $u_2 = 4$. This could mean in a physical sense that a gravitational load of 6 is applied at node 2, and there is an unknown reaction from earth of R_2. We also imply by the notation adopted that $u_2 = 4$.

(N.B. At first sight, it might seem that, from statics, R_2 must equal -6. However, $6 + R_2$ is balanced by the sum of the elastic forces provided by the elements meeting at node 2, just as the applied load of 11 at node 1 is balanced by the sum of the elastic forces provided by the elements meeting at node 1).

To solve the equations in (2.14) we should simply forget (b), substitute for u_2 in (a) and (c) and solve for u_1 and u_3. However, we generally require the value R for ancillary structural design purposes, and so we adopt the following procedure. The first stage is almost unaltered and we simply eliminate u_1 from (b) and (c)

$$3u_1 + u_2 + 2u_3 = 11 \tag{2.15a}$$

$$0u_1 + \tfrac{2}{3}u_2 + \tfrac{1}{3}u_3 = \tfrac{7}{3} + R_2, u_2 = 4 \tag{2.15b}$$

$$0u_1 + \tfrac{1}{3}u_2 + \tfrac{2}{3}u_3 = \tfrac{8}{3}. \tag{2.15c}$$

The second stage is, however, quite unfamiliar. We eliminate u_2 from (a) and (c) as before, but this time by substituting the known value,

$$3u_1 + 0u_2 + 2u_3 = 7 \tag{2.16a}$$

$$0u_1 + \tfrac{2}{3}u_2 + \tfrac{1}{3}u_3 = \tfrac{7}{3} + R_2, u_2 = 4 \tag{2.16b}$$

$$0u_1 + 0u_2 + \tfrac{2}{3}u_3 = \tfrac{4}{3}. \tag{2.16c}$$

In this special case PIVOT could be permitted to be zero for we shall never divide by it.

The third stage is very much as it was before, and we simply eliminate u_3 from (b) and (c)

$$3u_1 + 0u_2 + 0u_3 = 3 \tag{2.17a}$$

$$0u_1 + \tfrac{2}{3}u_2 + 0u_3 = \tfrac{5}{3} + R_2, u_2 = 4 \qquad (2.17b)$$

$$0u_1 + 0u_2 + \tfrac{2}{3}u_3 = \tfrac{4}{3}. \qquad (2,17c)$$

Thus, by inspection, we see that $u_1 = 1$, $u_2 = 4$ as given, $u_3 = 2$ and $R_2 = 1$. We note that for equation (b) the unknown appears on the R.H.S.

2.5.4 Gaussian elimination

Gauss–Jordan reduction is simple but, unfortunately, it destroys the banded nature of finite element problems. It has, however, served as a convenient introduction to the technique almost universally used, namely Gaussian reduction. To demonstrate the method we use the following example where

$$u_1 + 2u_2 + u_3 + u_4 = 12 \qquad (2.18a)$$

$$2u_1 + 5u_2 + 2u_3 + u_4 = 15 + R_2, u_1 = 2 \qquad (2.18b)$$

$$u_1 + 2u_2 + 3u_3 + 2u_4 = 22 \qquad (2.18c)$$

$$u_1 + u_2 + 2u_3 + 2u_4 = 17 \qquad (2.18d)$$

Firstly we eliminate u_1 from (b), (c) and (d) by taking twice (a) from (b), (a) from (c) and (a) from (d), so that

$$u_1 + 2u_2 + u_3 + u_4 = 12 \qquad (2.19a)$$

$$0u_1 + u_2 + 0u_3 - u_4 = -9 + R_2, u = 2 \qquad (2.19b)$$

$$0u_1 + 0u_2 + 2u_3 + u_4 = 10 \qquad (2.19c)$$

$$0u_1 - u_2 + u_3 + u_4 = 5. \qquad (2.19d)$$

At this stage there is no difference between Gauss–Jordan and Gaussian reduction. However, from this point the process is different. We use an equation, or we substitute, only *below* the current or active equation. (N.B. If we are eliminating u_r, the rth equation is active). Here the active equation is (b) so we modify (b), (c) and (d) in order to eliminate the u_2 coefficient. Because u_2 is prescribed, we merely substitute. Because the coefficient of u_2 is already zero in (c), we only modify (d):

$$u_1 + 2u_2 + u_3 + u_4 = 12 \qquad (2.20a)$$

$$0u_1 + 0u_2 + 0u_3 - u_4 = -11 + R_2, u_2 = 2 \qquad (2.20b)$$

$$0u_1 + 0u_2 + 2u_3 + u_4 = 10 \qquad (2.20c)$$

$$0u_1 + 0u_2 + u_3 + u_4 = 7. \qquad (2.20d)$$

Finally we use (c) to remove the coefficient of u_3 in (d) so that

$$u_1 + 2u_2 + u_3 + u_4 = 12 \tag{2.21a}$$

$$0u_1 + 0u_2 + 0u_3 - u_4 = -11 + R_2, u_2 = 2 \tag{2.21b}$$

$$0u_1 + 0u_2 + 2u_3 + u_4 = 10 \tag{2.21c}$$

$$0u_1 + 0u_2 + 0u_3 + \tfrac{1}{2}u_4 = 2. \tag{2.21d}$$

We now have a set of equations which can be solved directly if we take them in reverse order. Starting with (d), we have $u_4 = 4$. Having obtained u_4, (c) gives $u_3 = 3$. Having obtained u_4 and u_3 and with u_2 prescribed, (b) gives $R_2 = 7$ immediately. Completing our "backsubstitution", we can substitute u_2, u_3 and u_4 in (a) to find $u_1 = 1$.

We now summarise in matrix form the above result:

$$\begin{bmatrix} 1 & 2 & 1 & 1 \\ 2 & 5 & 2 & 1 \\ 1 & 2 & 3 & 2 \\ 1 & 1 & 2 & 2 \end{bmatrix} \begin{bmatrix} u_1 \\ 2 \\ u_3 \\ u_4 \end{bmatrix} = \begin{bmatrix} 12 \\ 15 + R_2 \\ 22 \\ 17 \end{bmatrix}$$

is reduced to

$$\begin{bmatrix} 1 & 2 & 1 & 1 \\ 0 & 0 & 0 & -1 \\ 0 & 0 & 2 & 1 \\ 0 & 0 & 0 & \tfrac{1}{2} \end{bmatrix} \begin{bmatrix} u_1 \\ 2 \\ u_3 \\ u_4 \end{bmatrix} = \begin{bmatrix} 12 \\ -11 + R_2 \\ 10 \\ 2 \end{bmatrix}.$$

2.5.5 Subroutine GREDUC

Object: We now examine a Gaussian reduction routine which restates in FORTRAN the equation solving technique we have just considered. We again use the notation in the previous section.

Dictionary of variables names

ASLOD(MEQNS)	ASsembled LOaD vector
ASTIF(MEQNS,MEQNS)	Assembled global STIFfness matrix
IEQNS,NEQNS,MEQNS	Index, Number, Maximum of EQquatioNS
IFPRE(MEQNS)	Vector of parameter defining the fixity of a node 0—free; 1—fixed.
FIXED(MEQNS)	Vector of prescribed displacements (zero if not prescribed)
ICOLS	Index COLumn of Structural stiffness matrix
IROWS	Index ROW of Structural stiffness matrix

FACTR   Gaussian reduction FACToR
PIVOT   Diagonal term of variable which is currently
                                         being eliminated.

Input/Output diagram

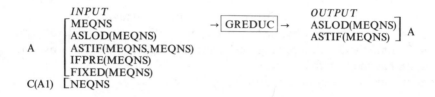

Annotated FORTRAN listing

```
          SUBROUTINE GREDUC(MEQNS,ASLOD,ASTIF,
        .IFPRE,FIXED)
   C
   C      GAUSSIAN REDUCTION ROUTINE
   C
          DIMENSION ASLOD(MEQNS),ASTIF(MEQNS,MEQNS)
          DIMENSION FIXED(MEQNS),IFPRE(MEQNS)
          COMMON/A1/ NEQNS,NEVAB,NELEM,NNODE,NDOFN
          DO 50 IEQNS=1,NEQNS                                    1*
          IF(IFPRE(IEQNS).EQ.1) GO TO 30                         2*
   C
   C      REDUCE EQUATIONS
   C
          PIVOT=ASTIF(IEQNS,IEQNS)                               3*
          IF(ABS(PIVOT).LT.1.0E-10) GO TO 60                     4*
          IF(IEQNS.EQ.NEQNS) GO TO 50
          IEQN1=IEQNS+1
          DO 20 IROWS=IEQN1,NEQNS
          FACTR=ASTIF(IROWS,IEQNS)/PIVOT
          IF(FACTR.EQ.0.0) GO TO 20
          DO 10 ICOLS=IEQNS,NEQNS
          ASTIF(IROWS,ICOLS)=ASTIF(IROWS,ICOLS)-                 5*
        .FACTR*ASTIF(IEQNS,ICOLS)
       10 CONTINUE
          ASLOD(IROWS)=ASLOD(IROWS)-
        .FACTR*ASLOD(IEQNS)
       20 CONTINUE
          GO TO 50
   C
   C      ADJUST RHS(LOADS) FOR
   C      PRESCRIBED DISPLACEMENTS
   C
```

```
      30 DO 40 IROWS=IEQNS,NEQNS
         ASLOD(IROWS)=ASLOD(IROWS)-
        .ASTIF(IROWS,IEQNS)*FIXED(IEQNS)
         ASTIF(IROWS,IEQNS)=0.0
      40 CONTINUE
         GO TO 50
      60 WRITE(6,100)
     100 FORMAT(5X,15HINCORRECT PIVOT)
         STOP
      50 CONTINUE
         RETURN
         END
```
$$\left.\begin{array}{l}\\\\\\\\\end{array}\right\}6*$$

1* Loop for each equation—this equation is associated with variable about to be eliminated
2* If this variable is fixed skip to 30
3* Extract PIVOT—the leading diagonal term
4* Check for zero or negative PIVOT in which case write a message and stop the program
5* Alter equations below equation 'IEQNS' not those above
6* For prescribed variables adjust the R.H.S.

General notes

For the problem described in Section 2.5.4. the main variables have the following values:

NEQNS = 4,

$$\text{ASLOD} = \begin{bmatrix} 12 \\ 15 \\ 22 \\ 7 \end{bmatrix} \quad \text{modified ASLOD} = \begin{bmatrix} 12 \\ -11 \\ 10 \\ 2 \end{bmatrix},$$

$$\text{ASTIF} = \begin{bmatrix} 1 & 2 & 1 & 1 \\ 2 & 5 & 2 & 1 \\ 1 & 2 & 3 & 2 \\ 1 & 1 & 2 & 2 \end{bmatrix}, \quad \text{modified ASTIF} = \begin{bmatrix} 1 & 2 & 1 & 1 \\ 0 & 0 & 0 & -1 \\ 0 & 0 & 2 & 1 \\ 0 & 0 & 0 & \frac{1}{2} \end{bmatrix},$$

$$\text{IFPRE} = \begin{bmatrix} 0 \\ 1 \\ 0 \\ 0 \end{bmatrix}, \quad \text{FIXED} = \begin{bmatrix} 0 \\ 2 \\ 0 \\ 0 \end{bmatrix}.$$

This subroutine is larger than JORDAN, but the work done is less. It will be further halved when we learn to use the symmetry of all the matrices. As with JORDAN, ASTIF and ASLOD undergo a series of transformations as we eliminate each variable in turn.

2.5.6 Subroutine BAKSUB

Object: Having completed the Gaussian reduction procedure we must now 'backsubstitute' to calculate the unknowns. We present a routine to perform this task.

Dictionary of variable names

ASLOD(MEQNS)	Reduced load vector
ASTIF(MEQNS,MEQNS)	Reduced global stiffness matrix
IEQNS,NEQNS,MEQNS	Index, Number, Maximum of EQuatioNS
IFPRE(MEQNS)	Vector of parameters defining the fixing of a node 0—free; 1—fixed
FIXED(MEQNS)	Vector of prescribed displacements (zero if not prescribed)
PIVOT	Diagonal term of variable currently being evaluated
REACT(MEQNS)	REACTions at nodes with prescribed displacements
XDISP(MEQNS)	Displacements at nodes

Input/Output diagram

```
          INPUT                                    OUTPUT
        ⎡ MEQNS              →⎡BAKSUB⎤→            XDISP(MEQNS) ⎤ A
        ⎢ ASLOD(MEQNS)                             REACT(MEQNS) ⎦
   A    ⎢ ASTIF(MEQNS,MEQNS)
        ⎢ IFPRE(MEQNS)
        ⎣ FIXED(MEQNS)
  C(A1) ⎣ NEQNS
```

Annotated FORTRAN listing

```
          SUBROUTINE BAKSUB(MEQNS,ASLOD,ASTIF,
         .IFPRE,FIXED,XDISP,REACT)
   C
   C      BACK-SUBSTITUTION ROUTINE
   C
          DIMENSION ASLOD(MEQNS),ASTIF(MEQNS,MEQNS)
          DIMENSION IFPRE(MEQNS),FIXED(MEQNS)
          DIMENSION XDISP(MEQNS),REACT(MEQNS)
          COMMON/A1/ NEQNS,NEVAB,NELEM,NNODE,NDOFN
          DO 5 IEQNS=1,NEQNS                          ⎫
          REACT(IEQNS)=0.0                            ⎬ 1*
        5 CONTINUE                                    ⎭
          NEQN1=NEQNS+1
```

```
        ┌──── DO 30 IEQNS=1,NEQNS                              ⎫
        │     NBACK=NEQN1-IEQNS                                ⎬ 2*
        │     PIVOT=ASTIF(NBACK,NBACK)                         ⎭ 3*
        │     RESID=ASLOD(NBACK)
        │     IF(NBACK.EQ.NEQNS) GO TO 20                        4*
        │     NBAC1=NBACK+1
        │  ┌── DO 10 ICOLS=NBAC1,NEQNS                         ⎫
        │  │   RESID=RESID-ASTIF(NBACK,ICOLS)*XDISP(ICOLS)     ⎬ 5*
        │  └ 10 CONTINUE                                        ⎭
        │     20 IF(IFPRE(NBACK).EQ.0)                         ⎫
        │       .XDISP(NBACK)=RESID/PIVOT                       ⎬ 6*
        │        IF(IFPRE(NBACK).EQ.1)                          ⎫
        │       .XDISP(NBACK)=FIXED(NBACK)                      ⎬ 7*
        │        IF(IFPRE(NBACK).EQ.1) REACT(NBACK)=-RESID     ⎭
        └── 30 CONTINUE
              RETURN
              END
```

1* Zero space for reactions
2* Loop backwards over each equation
3* Use the same PIVOT as in GREDUC
4* For last equation (the first to be solved) we do not have any other variables to substitute (i.e. bypass the loop)
5* Evaluate RESID from previously calculated variables
6* If variable not prescribed evaluate variable
7* If variable prescribed evaluate R.H.S. reaction

General notes

For the problem described in Section 2.5.4, the main variables have the following values:

NEQNS = 4,

$$ASLOD = \begin{bmatrix} 12 \\ -11 \\ 10 \\ 2 \end{bmatrix}, \quad ASTIF = \begin{bmatrix} 1 & 2 & 1 & 1 \\ 0 & 0 & 0 & -1 \\ 0 & 0 & 2 & 1 \\ 0 & 0 & 0 & \frac{1}{2} \end{bmatrix},$$

$$IFPRE = \begin{bmatrix} 0 \\ 1 \\ 0 \\ 0 \end{bmatrix}, \quad FIXED = \begin{bmatrix} 0 \\ 2 \\ 0 \\ 0 \end{bmatrix},$$

$$XDISP = \begin{bmatrix} 1 \\ 2 \\ 3 \\ 4 \end{bmatrix}, \quad REACT = \begin{bmatrix} 0 \\ 7 \\ 0 \\ 0 \end{bmatrix}$$

2.6 Symmetric matrices and banded matrices

In this section we describe certain special features of the stiffness equations in structural analysis problems which allow us to obtain economies in the computation. The two main features are:

(i) The symmetry of the global stiffness matrix

(ii) The banded nature of the global stiffness matrix.

The symmetry allows us to process only these terms on and above the leading diagonal of the global stiffness matrix. Matrix bandedness allows us to manipulate only a small portion of the global stiffness matrix as we solve for each variable.

2.6.1 Symmetry

With the Gaussian reduction procedure, when a variable such as u_s is eliminated

$$K_{ij} \text{ becomes } K_{ij} - K_{is}K_{sj}/K_{ss}$$

and

$$K_{ji} \text{ becomes } K_{ji} - K_{js}K_{si}/K_{ss}.$$

If $K_{ij} = K_{ji}$ (before elimination), $K_{is} = K_{si}$ and $K_{sj} = K_{js}$ then $K_{ij} = K_{ji}$ after the appropriate adjustment term has been subtracted. Indeed, the matrix remains symmetric throughout the elimination procedure. When we omitted all the terms below the diagonal as in (2.7) and (2.8), we were hinting how computer storage might be saved since only half of the matrix is required.

2.6.2 Banded equations

In fact, for most real problems, we need much less than half of the matrix. Equation (2.6) already exhibits recognisable 'banding' and it is usually more evident, the larger the problem. A 'banded' matrix has zero coefficients outside a band above and below the leading diagonal. Even within the band envelope zero coefficients usually exist. Figure 2.4 shows a typical banded matrix with a maximum semi-bandwidth M. In the so-called banded solutions, the upper triangular portion of a sub-matrix of dimensions $M \times M$ must be stored in core at any stage of the reduction process. Thus the storage requirements are $\frac{1}{2}M(M + 1)$. The size of M depends directly on how the nodes are numbered and consequently nodal numbering is very important with banded solutions.

Let us now enquire how "banding" would be evident in the array LNODS

—the array of node numbers defining each element described earlier in Section 2.4. Suppose a typical element has node numbers 27, 24 and 28. Its stiffness contributions are scattered to the nine terms in the global matrix: $K_{27,27}, K_{27,24}, K_{27,28}; K_{24,24}, K_{24,27}, K_{24,28}$ and $K_{28,27}, K_{28,24}, K_{28,28}$. None of these terms is far from the diagonal: the furthest is $K_{24,28}$. This is just four steps away from the diagonal in a horizontal or vertical direction.

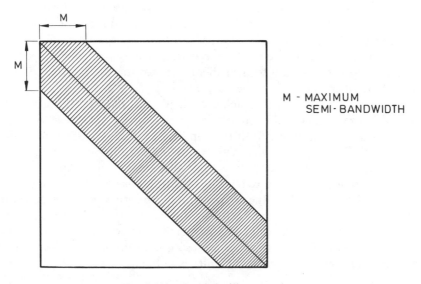

M - MAXIMUM
SEMI-BANDWIDTH

FIG. 2.4(a).   Banded stiffness matrix.

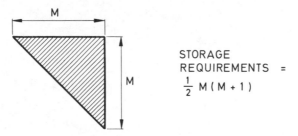

STORAGE
REQUIREMENTS =
$\frac{1}{2}$ M (M + 1)

FIG. 2.4(b).   Typical submatrix held in computer storage.

The 'semi-bandwidth' generated by this element is $(28-24)+1 = 5$. If another element had node numbers 31, 26 and 27, the semi-bandwidth would have to be increased to $6 = (31-26)+1$. The element with the greatest spread of node numbers evidently dictates the semi-bandwidth we must cater for, i.e. M. In general terms

$$M = (D + 1) \times \text{(number of degrees of freedom per node)},$$

where D is the maximum difference between any two node numbers occurring in any single element in the finite element mesh.

People who habitually use banded solutions soon develop the skill of choosing node numbers in such a way as to reduce M to nearly its minimum value. This correspondingly reduces the storage requirement. However, with elements having midside nodes or with large 3D problems, bandwidth minimisation is difficult so that experienced analysts often use specially written computer programs to help them re-order the nodes. Alternatively, they adopt other solution schemes.

Significant economy can be achieved if we can reduce calculations by recognising zeros included within the band envelope. In Chapter 8 we shall present a frontal solution FRONT again based on the Gaussian reduction process. FRONT requires fewer calculations and less computer storage space than banded solution since space is only allocated when required by non-zero row coefficients in the reduction process. Most of the programming effort in FRONT, and in the more complex solution schemes currently used in an industrial context, centres on a relatively sophisticated house keeping system. The main aim of Chapter 8 is therefore to persuade the reader that assembly and solution are essentially simple processes and in commercial programs the main difficulty is in the housekeeping.

3

Input and Output

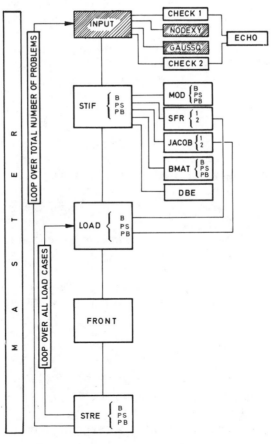

3.1 Introduction

Although the input of necessary data to and output of final results from any finite element program is one of the technically less demanding aspects, it is important to engineers. When the finite element method is employed in engineering analysis, the time spent in data preparation and interpretation of the eventual results generally represents the major proportion of the total time. Therefore any savings that can be made in these directions will be of direct benefit to the user.

The introduction of the isoparametric element concept has influenced the input–output situation considerably. The use of complicated elements such as those of the isoparametric family, generally means that fewer elements are required to give the same degree of accuracy than if simpler, constant stress elements were employed. As a rough guide a single parabolic isoparametric element, as shown in Fig. 1.5 can often be used in place of approximately 10 triangular constant stress elements; this of course depends on the problem geometry and the nature of the loading. Consequently the use of isoparametric elements generally results in a reduction in data preparation. However, in situations where fine geometrical detail has to be modelled, as is often the case in the aeronautical industry, the use of simple elements may still be unavoidable.

With regard to output of results the introduction of complex elements produces difficulties. In structural analysis by the displacement method the displacements at nodal points are the prime variables with the stress field being determined as a by-product. The nodal displacements are calculated and output in the usual way. However special considerations have to be given to the determination and output of stresses. It is in this situation that the relatively large area covered by a single isoparametric element appears at first sight to be disadvantageous.

The nodal points are generally concentrated along the boundaries of isoparametric elements, and therefore output of nodal stresses only is not satisfactory. It is desirable that stress values be produced at points inside the element, and for convenience, the Gaussian sampling points employed in the numerical integration of element stiffnesses and load terms are often utilised for this purpose.

3.2 Input data

For any finite element analysis the input data required can be subdivided into three main classifications. Firstly the data required to define the geometry of the structure and the support conditions must be furnished. Secondly, information regarding the material properties of the constituent materials must be prescribed. The final category concerns the loading to which the structure is subjected. In view of the wide variety of loading which may act (e.g. pressure, gravity, centrifugal, etc.) this subject will not be dealt with here, and Chapter 7 is devoted entirely to the determination of equivalent nodal forces.

Obviously a consistent set of units must be employed for all input data. Provided that all length and force terms are input in the same respective units, then the resulting displacements and stresses will be similarly dimensioned.

Details of the input data required are given in the following sections and detailed "user instructions" for the three programs developed in this text are included in Appendix I.

3.3 Control data

As previously indicated in Chapter 1, programs for three applications will be presented in this text. In order to reduce the total number of subroutines required several subroutines are structured so that they can be utilised in more than one program. To achieve this it is essential that some control information be supplied as input data. For example, the number of degrees of freedom per node will differ between plane stress/strain and plate bending applications. To enable the same data input subroutines to be used for both applications this information must be specified.

Use of variables in place of specific numerical values is also beneficial for understanding of the program, since if the control variables are suitably named their use and meaning are self explanatory to a large degree. For example, an improvement in clarity is immediately obtained by employing variable names for the number of coordinates per node and the number of degrees of freedom per node in coding, since for plane problems both quantities have the value 2.

A list of the control parameters required as input in order to allow the same input subroutine to be used in all three programs is now presented.

NPOIN	Total number of nodal points in the structure
NELEM	Total number of elements in the structure
NVFIX	Total number of boundary points, i.e. nodal points at which one or more degrees of freedom are restrained. It should be noted that in this context an internal node can be a boundary node.
NCASE	The total number of load cases to be solved for. Provided that the structural geometry remains unchanged the element

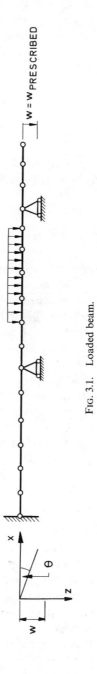

Fig. 3.1. Loaded beam.

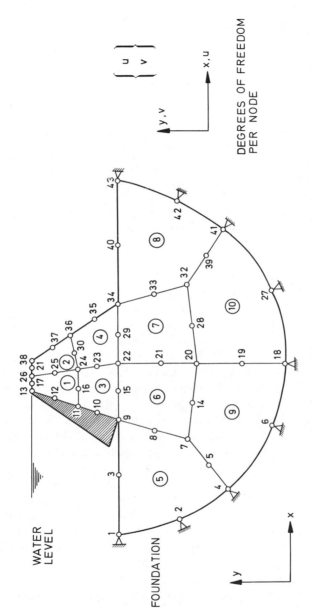

Fig. 3.2.   A typical two dimensional solid application gravity dam section

stiffnesses need not be recomputed for each additional loading case.

NTYPE   Problem type parameter   1—Plane stress
                                 2—Plane strain
This parameter is not applicable for beam and plate bending applications. For these cases enter a zero value.

NNODE   Number of nodes per element 3 for the beam program
                                    8 for plane stress/strain
                                    8 for plate bending

NDOFN   The number of degrees of freedom per nodal point. For a beam analysis this is 2; (w, θ), see Fig. 3.1, for two dimensional plane solids it is also 2; (u, v), Fig. 3.2, and for plate bending problems it is 3 (w, θ_x, θ_y), Fig. 3.3.

NMATS   Total number of different materials in the structure

NPROP   The number of material parameters required to define the characteristics of a material completely
3 Beam analysis
5 Plane Stress/strain
4 Plate bending
These are discussed in detail in Section 3.6

NGAUS   The order of Gaussian numerical integration rule to be employed. For isoparametric elements the use of numerical techniques for volume integration, etc., is unavoidable and a Gauss quadrature rule is employed in this program. The order of rule is left as a user option since the optimum value varies with circumstances. Generally for isoparametric elements with quadratic variation a three point Gauss rule for each dimension is considered the optimum. However, in situations where initial strains are present (e.g. thermal stresses) or the aspect ratio of elements is allowed to become high it has recently become apparent [1] that use of a reduced integration order is beneficial. In such situations a two point Gauss rule can be employed with parabolic elements. Thus if NGAUS is prescribed as 2 a two-point Gauss rule is to be employed; if NGAUS is input as 3 a three-point rule will be used. Numerical integration procedures will be considered in more detail in Chapter 4.

NDIME   Number of coordinate components required to define each nodal point
1 Beam analysis
2 Plane stress/strain

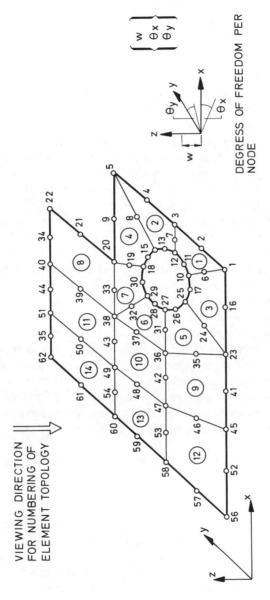

FIG. 3.3. Plate bending application

2 Plate bending

NSTRE   Number of independent stress components at any point

2 Beam program

3 Plane stress/strain

5 Plate bending

3.4 Geometric data

Once the structure to be analysed has been discretised into a number of finite elements, the structural geometry must then be defined numerically. Each node is first identified by prescribing a number of each point as illustrated in Fig. 3.2. If a method of equation solution which relies on an overall banded matrix being employed were to be used, then the nodal points must be numbered in a systematic manner. Each element must also be designated an identification number, an arbitrary sequence being suitable in this case.

However, if the frontal method of solution is to be employed, the nodal points can be numbered in a random manner if necessary. This is a major advantage, since, if a node number is neglected during the initial numbering process, then it can be subsequently allocated a value without necessitating the renumbering of the other points as would probably be demanded by a banded solution technique. For frontal solution, it is the order in which the elements are processed that is of importance; an attempt being necessary to optimise this order so as to minimise the frontwidth. Therefore it is convenient to number the elements in the order in which it is intended to process them. When a structure is analysed by parabolic isoparametric elements, the total number of elements is always far less than the total number of nodal points. The task of element renumbering for front width optimisation is then considerably easier than nodal point renumbering for bandwidth minimisation. In the following programs the frontal method will be adopted for solution.

The geometry of the structure can then be completely defined by prescribing the following two sets of information.

(a) Specification of element connections
The geometry of each individual element must be specified by listing in a systematic way the numbers of the nodal points which define its outline. Each element is identified by its element number and this, in the frontal solution method, also indicates the order in which the elements are processed for stiffness formulation and assembly. The element topology is read into the array

LNODS(NUMEL,INODE),

where NUMEL corresponds to the number of the element under consideration, and subscript INODE ranges from 1 to NNODE thus designating the
NNODE nodal point numbers defining the element. Since, in principle, it is
possible to assign different material properties to each element, a material
property identification number is also allocated to each element. This is
achieved by means of the array

MATNO(NUMEL),

meaning that element number NUMEL has material properties of type
MATNO(NUMEL) which have to be input at another stage. Therefore in
order to define the topology and material properties of each element, the
following information must be input.

NUMEL,MATNO(NUMEL),

(LNODS(NUMEL,INODE),INODE = 1,NNODE).

This information is required for NUMEL = 1 to NELEM where NELEM
is the total number of elements.

The specification of the nodal connection numbers must follow a systematic
pattern. In this text we follow the convention that an anticlockwise sequence
is adhered to, beginning from any corner node. Thus for the two-dimensional
element 3 in Fig. 3.2 the nodal connections would be specified as

9   15   22   23   24   16   11   10

or,

22   23   24   16   11   10   9   15

etc.

For beam analysis the three element nodal connection numbers are input
in sequence starting from any end node and the orientation of the coordinate
axes employed is shown in Fig. 3.1.

For plane stress and strain problems the relative orientation of the x and y
axes are as shown in Fig. 3.2 with the independent coordinate direction z
coming out of the paper. For plate bending applications the coordinate
system employed is oriented as shown in Fig. 3.3. For listing the nodal
connection numbers the elements must be viewed from above the plane $z = 0$.

(b) Specification of the spatial coordinates of each nodal point
The coordinates of each nodal point must be defined with reference to a
global coordinate system. This information will be read in and stored in the
array

COORD(IPOIN,IDIME),

where IPOIN corresponds to the number of the nodal point and IDIME refers to the coordinate components.

For beam analysis the position of each nodal point is defined by a single coordinate component, usually the horizontal distance measured from some arbitrary datum. For both plane stress/strain and plate bending problems two coordinate components are required and a Cartesian reference system (x, y) will be adopted with the origin being arbitrarily chosen. Thus for nodal point IPOIN the following information must be input

$$\text{IPOIN,(COORD(IPOIN,IDIME),IDIME} = 1,\text{NDIME),}$$

where NDIME is the total number of coordinate dimensions as defined in Section 3.3. For plane and plate situations each element side is defined by three nodal points as illustrated in Fig. 1.5. If an element side is linear then clearly it requires only the coordinate of two nodal points to define its position. Therefore in order to reduce the amount of data preparation the input subroutine will be structured so that it is unnecessary to input the coordinates of a mid-side node if it lies on a straight line connecting the two adjacent corner nodes.

3.5 Boundary conditions

With the geometry of the structure defined it is now necessary to specify the boundary conditions. In the displacement method these take the form of prescribed values of the relevant degrees of freedom at nodal points. If values of the traction components are to be prescribed on any part of the structure, then these must be input as equivalent nodal forces.

Since the number of degrees of freedom per node will vary between beam analysis, two dimensional solids and plate bending, the input data for restrained nodal points will differ; although the same structure of input will be followed in each case. The nodes at which one or more degrees of freedom are restrained are read into the array

$$\text{NOFIX(IVFIX),}$$

which signifies that the IVFIXth boundary node to be specified has a nodal point number NOFIX(IVFIX).

A means must now be introduced of specifying which degrees of freedom at a node are to be restrained or prescribed with specified displacement values. This can be achieved by use of an integer code input via the array

$$\text{IFPRE(IVFIX,IDOFN),}$$

where IDOFN ranges over the number of degrees of freedom per node,

NDOFN. This states that the IDOFNth degree of freedom of the IVFIXth boundary node has a fixity code value of IFPRE(IVFIX,IDOFN). For *beam analysis* the code value IFPRE(IVFIX,IDOFN) may have the following values.

   10   Displacement w in the z direction restrained
   01   Rotation component θ restrained
   11   Both the w displacement and rotation θ restrained

The directions of w and θ are defined in Fig. 3.1.

For *two-dimensional solid* applications code value IFPRE(IVFIX,IDOFN) may have three values as defined below.

   10   Displacement in the x direction restrained
   01   Displacement in the y direction restrained
   11   Displacements in both x and y directions restrained

The x and y directions are defined in Fig. 3.2.

For *plate bending* applications code value IFPRE(IVFIX,IDOFN) may have the following values

   100   Displacement, w, normal to the plane of the plate restrained
   010   Rotation component θ_x restrained
   001   Rotation component θ_y restrained
   110   Normal displacement and rotation component θ_x restrained
          etc.

The directions of w, θ_x and θ_y are defined in Fig. 3.3. In other words a unit value in the relevant column indicates a fixed degree of freedom whereas a zero entry indicates no restraint of that particular component. A prescribed value of any degree of freedom (displacement or rotation) is input into the following array

$$PRESC(IVFIX,IDOFN),$$

where IVFIX indicates that the prescribed displacements pertain to the IVFIXth boundary node and IDOFN ranges over the number of degrees of freedom per node.

Thus the data relating to a restrained node is contained on one card as follows

$$NOFIX(IVFIX),(IFRE(IVFIX,IDOFN),IDOFN=1,NDOFN),$$

$$(PRESC(IVFIX,IDOFN),IDOFN=1,NDOFN).$$

This information is input for IVFIX = 1 to NVFIX where NVFIX is the total number of restrained nodes. It should be observed that a fixed boundary point is regarded as a node at which zero displacement values are prescribed.

3.6 Material properties

The material properties required for solution will differ for the three applications, but as far as input of the necessary data is concerned, the same storage array will be employed in each case.

Each different material is allocated an identification number and the properties relating to this goes into the array

$$PROPS(NUMAT,IPROP),$$

where NUMAT denotes the material identification number and the subscript IPROP the individual property. Each element is associated with a particular material type through the previously mentioned identification array MATNO(NUMEL).

The required properties for beam analysis will be discussed in detail in Chapter 4. The element behaviour is completely defined by specification of the flexural rigidity, EI, and the shear constant $5GA/6$ where E is the elastic modulus and A is the beam crosssectional area. Any distributed load per unit length of beam is treated as an additional material property.

For both the two-dimensional plane analysis and plate bending, the elastic material properties will be characterised by two independent material constants. These are chosen to be the elastic modulus, E, and Poisson's ratio, v. Since in both applications the thickness normal to the xy plane can vary from element to element it is convenient to treat this as an additional material property. This allows cases such as stepped plates to be analysed, but it does not permit smooth transitions in thickness throughout the structure. If this latter facility is required then the thickness must be prescribed as a nodal quantity. Such a treatment is not included here.

In the plane stress/strain analysis program the stresses and displacements induced by gravity will be included together with thermal stress effects produced by a specific temperature field. Consequently, for this case, the material density ρ, and the coefficient of thermal expansion α, must also be input. As for the beam, in plate bending applications it is convenient to regard any uniformly distributed load acting on an element as an additional material property. Since in many applications a constant uniformly distributed load may act over large areas of the structure, (e.g. self weight) this results in savings in data preparation.

Thus for each different material the following information must be input

$$NUMAT,(PROPS(NUMAT,IPROP),IPROP=1,NPROP),$$

where NUMAT is the material property identification number and PROPS (NUMAT,IPROP) are as listed below for each application. Index IPROP ranges over the total number of independent material properties, NPROP.

(*a*) *Beam*

PROPS(NUMAT,1)—Flexural rigidity, EI
PROPS(NUMAT,2)—Shear Constant, $GA/1 \cdot 2$
PROPS(NUMAT,3)—Distributed load intensity

(*b*) *Plane stress/strain*

PROPS(NUMAT,1)—Elastic modulus, E
PROPS(NUMAT,2)—Poissons ratio, v
PROPS(NUMAT,3)—Material thickness, t
PROPS(NUMAT,4)—Mass density, ρ
PROPS(NUMAT,5)—Coefficient of thermal expansion, α

(*c*) *Plate bending*

PROPS(NUMAT,1)—Elasticity modulus, E
PROPS(NUMAT,2)—Poissons ratio, v
PROPS(NUMAT,3)—Material thickness, t
PROPS(NUMAT,4)—Intensity of any uniformly distributed load

Regions in which one or more of these quantities differ must be treated as different materials.

3.7 Subroutine INPUT

This subroutine should be self-explanatory once the preceding sections have been read. Subroutine INPUT inputs the control data, geometric data, boundary conditions and material properties. In addition it calls subroutine NODEXY whose function is to generate the coordinates of midside nodes which lie on a straight line connecting adjacent corner nodes. It also calls subroutine GAUSSQ whose function is to generate the sampling point positions and weighting factors according to the order of integration rule specified, through NGAUS, in the control data. In order to enable the same subroutine to be employed in all three applications, a branch is included whereby separate statements are employed to accept boundary data for applications in which there are 2 degrees of freedom per node (beam and plane stress/strain problems) and for the case when 3 degrees of freedom per node exist (plate bending).

Subroutine INPUT is now listed.

Input/Output diagram

	INPUT		OUTPUT	
	NPOIN	→ INPUT →	NPOIN	
	NELEM		NELEM	
	NVFIX		NVFIX	
	NCASE		NCASE	
	NTYPE		NTYPE	
F(5)	NNODE		NNODE	C(CONTRO), F(6)
	NDOFN		NDOFN	
	NMATS		NMATS	
	NPROP		NPROP	
	NGAUS		NGAUS	
	NDIME		NDIME	
	NSTRE		NSTRE	

	LNODS(NELEM,NNODE)	LNODS(NELEM,NNODE)	
	MATNO(NELEM)	MATNO(NELEM)	
	COORD(NPOIN,NDIME)	COORD(NPOIN,NDIME)	
C(LGDATA)	NOFIX(NVFIX)	NOFIX(NVFIX)	C(LGDATA),
	IFPRE(NVFIX,NDOFN)	IFPRE(NVFIX,NDOFN)	F(6)
	PRESC(NVFIX,NDOFN)	PRESC(NVFIX,NDOFN)	
	PROPS(NMATS,NPROP)	PROPS(NMATS,NPROP)	

Annotated FORTRAN *listing*

```
        SUBROUTINE INPUT

      ┌─────────────────────┐
      │  COMMON  BLOCKS     │
      └─────────────────────┘

C
C*** READ THE FIRST DATA CARD, AND ECHO IT
C     IMMEDIATELY.
C
      READ(5,900) NPOIN,NELEM,NVFIX,NCASE,NTYPE,
     . NNODE,NDOFN,NMATS,NPROP,NGAUS,NDIME,NSTRE
  900 FORMAT(12I5)
      NEVAB=NDOFN*NNODE
      WRITE(6,905) NPOIN,NELEM,NVFIX,NCASE,NTYPE,
     . NNODE,NDOFN,NMATS,NPROP,NGAUS,NDIME,
     . NSTRE,NEVAB
  905 FORMAT(//8H NPOIN =,I4,4X,8H NELEM =,I4,
     . 4X,8H NVFIX =,I4,4X,8H NCASE =,I4,4X,
     . 8H NTYPE =,I4,4X,8H NNODE =,I4,4X,
     . 8H NDOFN =,I4// 8H NMATS =,I4,4X,
     . 8H NPROP =,I4,4X,8H NGAUS =,I4,4X,
     . 8H NDIME =,I4,4X,8H NSTRE =,I4,4X,
     . 8H NEVAB =,I4)
      CALL CHECK1
C
C*** READ THE ELEMENT NODAL CONNECTIONS, AND
C     THE PROPERTY NUMBERS.
C
```

```
          WRITE(6,910)
      910 FORMAT(//8H ELEMENT,3X,8HPROPERTY,6X,
         . 12HNODE NUMBERS)
          DO 10 IELEM=1,NELEM
          READ(5,900) NUMEL,MATNO(NUMEL),
         . (LNODS(NUMEL,INODE),INODE=1,NNODE)
       10 WRITE(6,915) NUMEL,MATNO(NUMEL),
         . (LNODS(NUMEL,INODE),INODE=1,NNODE)
      915 FORMAT(1X,I5,I9,6X,8I5)
C
C*** ZERO ALL THE NODAL COORDINATES, PRIOR
C    TO READING SOME OF THEM.
C
          DO 20 IPOIN=1,NPOIN
          DO 20 IDIME=1,NDIME
       20 COORD(IPOIN,IDIME)=0.0
C
C*** READ SOME NODAL COORDINATES, FINISHING
C    WITH THE LAST NODE OF ALL.
C
          WRITE(6,920)
      920 FORMAT(//25H   NODAL POINT COORDINATES)
          WRITE(6,925)
      925 FORMAT(6H   NODE,7X,1HX,9X,1HY)
       30 READ(5,930) IPOIN,(COORD(IPOIN,IDIME),
         . IDIME=1,NDIME)
      930 FORMAT(I5,5F10.5)
          IF(IPOIN.NE.NPOIN) GO TO 30
C
C*** INTERPOLATE COORDINATES OF MID-SIDE NODES
C
          IF(NDIME.EQ.1) GO TO 40
          CALL NODEXY
       40 CONTINUE
          DO 50 IPOIN=1,NPOIN
       50 WRITE(6,935) IPOIN,(COORD(IPOIN,IDIME),
         . IDIME=1,NDIME)
      935 FORMAT(1X,I5,3F10.3)
C
C*** READ THE FIXED VALUES.
C
          WRITE(6,940)
      940 FORMAT(//17H RESTRAINED NODES)
          WRITE(6,945)
      945 FORMAT(5H NODE,1X,4HCODE,6X,
         . 12HFIXED VALUES)
          IF(NDOFN.NE.2) GO TO 70
          DO 60 IVFIX=1,NVFIX
          READ(5,950) NOFIX(IVFIX),(IFPRE(IVFIX,
         . IDOFN),IDOFN=1,NDOFN),(PRESC(IVFIX,IDOFN),
         . IDOFN=1,NDOFN)
       60 WRITE(6,950) NOFIX(IVFIX),(IFPRE(IVFIX,
         . IDOFN),IDOFN=1,NDOFN),(PRESC(IVFIX,IDOFN),
         . IDOFN=1,NDOFN)
```

```
  950 FORMAT(1X,I4,3X,2I1,2F10,6)
      GO TO 90
   70 DO 80 IVFIX=1,NVFIX
      READ(5,955) NOFIX(IVFIX),(IFPRE(IVFIX,
     . IDOFN),IDOFN=1,NDOFN),(PRESC(IVFIX,
     . IDOFN),IDOFN=1,NDOFN)
   80 WRITE(6,955) NOFIX(IVFIX),(IFPRE(IVFIX,
     . IDOFN),IDOFN=1,NDOFN),(PRESC(IVFIX,
     . IDOFN),IDOFN=1,NDOFN)
  955 FORMAT(1X,I4,2X,3I1,3F10,6)
   90 CONTINUE
C
C*** READ THE AVAILABLE SELECTION OF ELEMENT
C    PROPERTIES,
C
      WRITE(6,960)
  960 FORMAT(//21H  MATERIAL PROPERTIES)
      WRITE(6,965)
  965 FORMAT(8H  NUMBER,7X,10HPROPERTIES)
      DO 100 IMATS=1,NMATS
      READ(5,930) NUMAT,(PROPS(NUMAT,IPROP),
     . IPROP=1,NPROP)
  100 WRITE(6,970) NUMAT,(PROPS(NUMAT,IPROP),
     . IPROP=1,NPROP)
  970 FORMAT(1X,I5,7X,5F14,6)
C
C*** SET UP GAUSSIAN INTEGRATION CONSTANTS
C
      CALL GAUSSQ
      CALL CHECK2
      RETURN
      END
```

One point worthy of note arises in the input of the nodal point coordinates. If a midside node is colinear with the two adjacent corner nodes then its coordinates need not be input, as they will be generated by subroutine NODEXY. Consequently the exact number of nodal points whose coordinates are to be input will generally be less than the total number of nodal points NPOIN. To avoid counting the number of nodal point coordinates to be read, we insist that the coordinates of the highest numbered node be input last whether it is a midside node or not. As soon as this card is encountered the program assumes that the input of nodal point coordinates has been completed and proceeds to the definition of element topologies.

It is seen that as soon as the control parameters have been input and printed subroutine CHECK1 is called to check this data. After the remaining data has been input it is checked by subroutine CHECK2 which is called before leaving subroutine INPUT. These error diagnostic subroutines are developed in Chapter 9.

3.8 Generation of coordinate values for midside nodes

Since the nodal point coordinate array COORD(IPOIN,IDIME) has been initially set to zero, any nodal point whose coordinates have not been directly input will retain these values. Subroutine NODEXY checks each midside node (a midside node being recognisable from the element topology cards). If both coordinates of a midside node are found to be zero, its coordinates are interpolated as being midway between the two adjacent corner nodes. This procedure introduces a minor complication if a particular midside node lies on a curve and is required to have zero coordinate values. In such a situation, to avoid complicating the program logic, the coordinates must be prescribed as very small non-zero values (say 10^{-10} units).

3.9 Subroutine NODEXY

This subroutine can be constructed and explanatory remarks provided.

Input/Output diagram

```
            INPUT                                    OUTPUT
            ┌                      ┌─────────┐
C(LGDATA)   │LNODS(NELEM,NNODE) → │ NODEXY  │ → Modified                    ┐
            │COORD(NPOIN,NDIME)   └─────────┘    COORD(NPOIN,NDIME)         │ C(LGDATA)
                                                                            ┘
            ┌                                      NELEM ┐
C(CONTRO)   │NELEM                                 NNODE │ C(CONTRO)
            │NNODE                                       ┘
```

Annotated FORTRAN listing

```
        SUBROUTINE NODEXY

  ┌─────────────────────┐
  │ COMMON  BLOCKS      │
  └─────────────────────┘

C
C*** LOOP OVER EACH ELEMENT
C
  ┌──── DO 30 IELEM=1,NELEM                             1*
  │ C
  │ C*** LOOP OVER EACH ELEMENT EDGE
  │ C
  │ ┌──── DO 20 INODE=1,NNODE,2                         2*
  │ │ C
  │ │ C*** COMPUTE THE NODE NUMBER OF THE FIRST NODE
  │ │ C
  │ │     NODST=LNODS(IELFM,INODE)          ⎫
  │ │     IGASH=INODE+2                     ⎬ 3*
  │ │     IF(IGASH.GT.NNODE) IGASH=1        ⎭
  │ │ C
```

```
C*** COMPUTE THE NODE NUMBER OF THE LAST NODE
C
      NODFN=LNODS(IELEM,IGASH)
      MIDPT=INODE+1
C
C*** COMPUTE THE NODE NUMBER OF THE
C    INTERMEDIATE NODE
C
      NODMD=LNODS(IELEM,MIDPT)
      TOTAL=ABS(COORD(NODMD,1))+                         } 4*
    . ABS(COORD(NODMD,2))
C
C*** IF THE COORDINATES OF THE INTERMEDIATE
C    NODE ARE BOTH ZERO INTERPOLATE BY A
C    STRAIGHT LINE
C
      IF(TOTAL.GT.0.0) GO TO 20                             5*
      KOUNT=1
   10 COORD(NODMD,KOUNT)=(COORD(NODST,KOUNT)+
    . COORD(NODFN,KOUNT))/2.0                           } 6*
      KOUNT=KOUNT+1
      IF(KOUNT.EQ.2) GO TO 10
   20 CONTINUE
   30 CONTINUE
      RETURN
      END
```

1* Loop over each element
2* Loop over each corner node of a particular element
3* Determine the nodal point numbers of two consecutive corner nodes
4* Determine the nodal point number of the midside node lying between
    the two corner nodes defined in 3* and compute the sum of the absolute
    values of its coordinates
5* Check if both coordinates of this midside node are zero. If not proceed
    to the next pair of consecutive corner nodes
6* If the midside node coordinates are zero, interpolate them as being
    mid-way between the two consecutive corner nodes. This is done for
    both the x and y coordinates.

3.10 Subroutine GAUSSQ

The function of this subroutine, which is called by subroutine INPUT, is to
set up the sampling point positions and weighting factors for numerical
integration. The Gauss quadrature routines utilised in this text are restricted
to either two- or three-point integration rules. The role of numerical integration
in the isoparametric formulation is discussed in detail in Chapter 4. The order
of integration rule to be employed is defined by NGAUS in the control

data input in subroutine INPUT. The sampling point positions are stored
in the array POSGP( ) while the weighting factors are stored in WEIGP( ).
Once again explanatory notes are provided after the subroutine listing.

Input/Output diagram

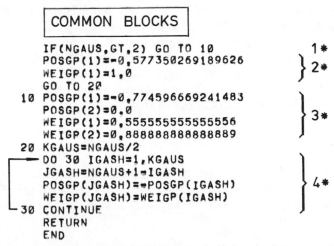

Annotated FORTRAN listing

```
        SUBROUTINE GAUSSQ

    ┌─────────────────────┐
    │  COMMON  BLOCKS     │
    └─────────────────────┘

        IF(NGAUS.GT.2) GO TO 10                          1*
        POSGP(1)=-0.577350269189626        ⎫
        WEIGP(1)=1.0                       ⎬ 2*
        GO TO 20                           ⎭
    10  POSGP(1)=-0.774596669241483        ⎫
        POSGP(2)=0.0                       ⎪
        WEIGP(1)=0.555555555555556         ⎬ 3*
        WEIGP(2)=0.888888888888889         ⎭
    20  KGAUS=NGAUS/2
    ┌──►DO 30 IGASH=1,KGAUS               ⎫
    │   JGASH=NGAUS+1-IGASH               ⎪
    │   POSGP(JGASH)=-POSGP(IGASH)        ⎬ 4*
    │   WEIGP(JGASH)=WEIGP(IGASH)         ⎭
    └─30 CONTINUE
        RETURN
        END
```

1* If NGAUS=2 continue. If NGAUS=3 go to 10.
2* Generate sampling point positions POSGP( ) and weighting factors
   WEIGP( ) for the two-point Gauss quadrature integration rule.
3* Generate similar constants for the three-point rule.
4* Compute the negative values by symmetry.

3.11 Presentation of output

At this stage all information, other than loading data, has been input and the
formation of element stiffnesses and solution of equations can proceed. The
nodal displacements (or displacement and rotations for plate bending) will
be calculated in the equation solution subroutine which is described in
detail in Chapter 8. It then remains to output these displacements and calculate

the stresses in each element. The stresses are obtained from the displacements through relationship (1.32). Since the **B** and **D** matrices (defined for plane problems by (1.17) and (1.20) respectively) have not yet been explicitly formulated it would be illogical to attempt to develop a stress output subroutine at this stage. The subroutine for the output of the Cartesian stress components at the Gaussian sampling points of each elements are formulated in Chapters 4, 5 and 6 for the beam, plane and plate bending applications respectively.

3.12 Automatic data preparation and output plotting

At this stage it is perhaps appropriate to mention briefly the aids which are available for preparing and checking input data and processing the final results. Since by far the greatest task in any finite element analysis is generally the preparation of the input data, and in particular definition of the nodal coordinates and element topology, any savings in effort that can be made in this area will be important. For this purpose mesh generation programs can be developed. These are generally of two types:

Where an electronic digitiser is employed to define and produce the geometric data

A semi-automatic approach where the structure is divided into a few large zones and the fineness of element subdivision within each is specified. The initial data is input in the normal way and the subdivision proceeds automatically

After the geometrical input data has been prepared, it is worthwhile to plot this automatically before attempting a finite element solution. Indeed, a graphical plot of the mesh offers a far better check on the geometric data than the use of error diagnostic subroutines. Since even if no data errors are detected by the diagnostic subroutines and a finite element solution is performed, it is still possible that the coordinate location of some nodal points may be incorrect and that the aspect ratio or distortion of some elements may be unacceptable for an accurate solution. If a sophisticated application (e.g. elasto-plastic stress analysis) is envisaged, a preliminary plot of the mesh can often result in large savings with respect to abortive runs.

The advent of mini-computers and the parallel development of interactive graphic systems promises to have a marked influence in the data preparation field. A preliminary mesh can first be generated by an electronic digitiser, and plotted on a visual display unit. Any errors, corrections or nodal coordinate adjustments can then be made directly by an interactive link. If a three-dimensional mesh is being prepared the viewing angle can even be repeatedly changed in order to obtain the most advantageous view of the structure. In this way data preparation and checking can be performed in one step.

Graphics programs can also be utilised in the processing of the final results. Plotting packages have been developed for plotting the deformed shape of structures, producing stress contours or principal stress vectors, etc. Such plots indicate to the engineer the areas where a closer examination of the stresses are necessary; the computer printout being employed at this stage. Interactive graphics systems are already having an impact in this area also, with programs being developed to allow the engineer to isolate and display critical regions of a structure and to vary the output quantity being plotted.

Ultimately it may be possible to dovetail the entire operation, with the data being generated and the results obtained and displayed in one operation, leading eventually to an interactive analysis/design process.

References

1. Zienkiewicz, O. C., Taylor, R. L., and Too, J. M., Reduced integration technique in general analysis of plates and shells. *Int. J. Num. Meth. Eng.* **3**, 275–290, 1971.

4

Isoparametric Beam Element

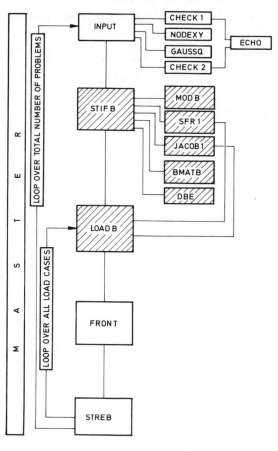

4.1 Introduction

In this chapter we examine the finite element formulation of a straight line element which can be used to analyse beams in which transverse shear deformation effects may be important. In keeping with the main philosophy of the book we choose a quadratic isoparametric element.

The main aim of the chapter is to use the simple example of the 1D beam element to introduce certain concepts which will be discussed later in greater detail for 2D elements. These concepts are then used in the development of subroutines to evaluate the stiffness and stress matrices and consistent load vectors for the beam element.

Unlike the more usual types of beam element, the isoparametric beam element can take account of transverse shear deformation since energy due to shear as well as bending is considered in the formulation. The element is quite versatile and can be used to analyse not only thin beams with negligible shear deformation but also thick beams and beams of sandwich construction in which shear effects are important. In this book, however, we shall only consider beams of homogeneous section.

The main assumption we make is concerned with the cross-sectional behaviour of the beam. Usually, in beam theory we assume that normals to the neutral axis before deformation remain straight and normal to the neutral axis after deformation. Indeed, if we took the trouble to paint a straight line representing the normal on the side of an unloaded thin beam, the line would remain straight and normal to the neutral axis after the beam was loaded as shown in Fig. 4.1(a). This would not be true for a thick beam. The normal would take up the shape indicated in Fig. 4.1(b) and could be approximately represented by a straight line inclined at an angle

$$\theta = \frac{\partial w}{\partial x} + \phi \tag{4.1}$$

to the x-axis, where $\partial w/\partial x$ is the slope of the neutral axis and ϕ is an extra rotation due to the transverse shear effects. Thus the total potential energy of the beam may be written as

$$\pi = \tfrac{1}{2} \int EI \left(\frac{\partial \theta}{\partial x} \right)^2 \mathrm{d}x \underbrace{\qquad}_{\text{flexural strain energy}} + \tfrac{1}{2} \int S\phi^2 \, \mathrm{d}x \underbrace{\qquad}_{\text{shear strain energy}} \tag{4.2}$$

$$\underbrace{- \int qw \, \mathrm{d}x}_{\substack{\text{work done by distributed} \\ \text{load } q}} \quad \underbrace{- \quad P.w}_{\substack{\text{work done by isolated} \\ \text{load } P}} \quad \underbrace{- \quad C.\theta}_{\substack{\text{work done by isolated} \\ \text{couple } C}}$$

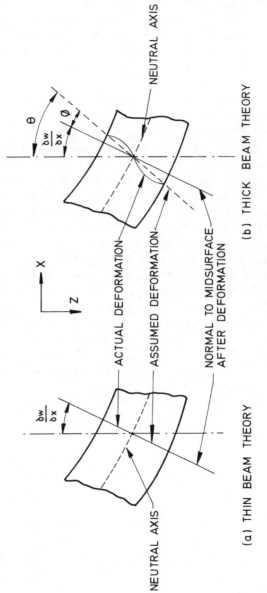

FIG. 4.1   Cross-sectional deformation of beam.

where EI = flexural rigidity = elastic modulus × second moment of area,

$$S = GA/\alpha = \text{shear rigidity} = \frac{\text{Shear modulus} \times \text{cross-sectional area}}{\text{factor to allow for warping}},$$

$\partial\theta/\partial x$ = pseudo-curvature,

ϕ = effective shear rotation.

We now describe the finite element formulation.

4.2 Element definition

This straight line beam element has three nodes—one at each end and one

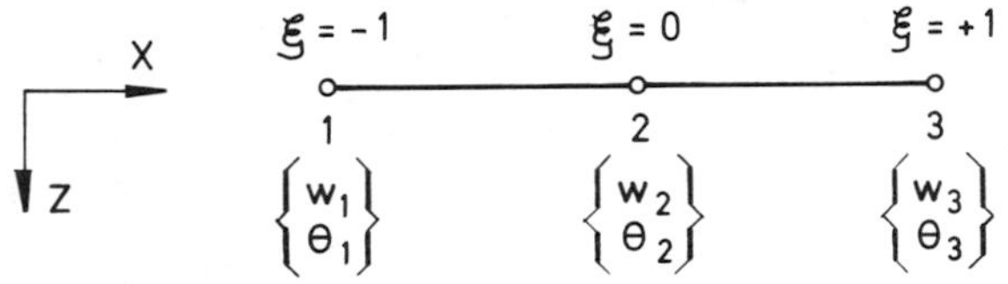

FIG. 4.2. Parabolic isoparametric thick beam element.

somewhere in between as shown in Fig. 4.2. We use two coordinate schemes in the element formulation:

the global coordinate system (x)

the natural coordinate system for the element (ξ)

at node 1 $\xi = -1$,

at node 2 $\xi = 0$ (if the node is halfway along the element),

at node 3 $\xi = +1$.

Each node i has two displacement degrees of freedom associated with it:

w_i the lateral displacement of the beam,

and

$$\theta_i = \left(\frac{\partial w}{\partial x}\right)_i + \phi_i \quad \text{the rotation of the normal.}$$

Thus, the element displacements may be listed in the vector

$$\delta^e = [w_1, \theta_1, w_2, \theta_2, w_3, \theta_3]^T. \tag{4.3}$$

A shape function or interpolation function is associated with each node as shown in Fig. 4.3.

$$N_1 = -\tfrac{1}{2}\xi(1 - \xi)$$
$$N_2 = (1 - \xi)(1 + \xi)$$
$$N_3 = \tfrac{1}{2}\xi(1 + \xi). \tag{4.4}$$

Thus, the lateral displacement $w(\xi)$ at any point within the element can be

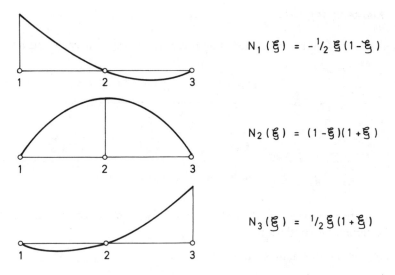

FIG. 4.3.   Parabolic shape functions for the beam element.

defined in terms of the shape functions and the associated nodal displacements by simple interpolation

$$w(\xi) = N_1(\xi)w_1 + N_2(\xi)w_2 + N_3(\xi)w_3 = \sum_{i=1}^{3} N_i w_i. \tag{4.5}$$

The quadratic interpolation scheme can be used to define the rotation $\theta(\xi)$ at any point within the element

$$\theta(\xi) = N_1(\xi)\theta_1 + N_2(\xi)\theta_2 + N_3(\xi)\theta_3. \tag{4.6}$$

The x-coordinate may be defined in this manner also

$$x(\xi) = N_1(\xi)x_1 + N_2(\xi)x_2 + N_3(\xi)x_3 = \sum_{i=1}^{3} N_i x_i. \tag{4.7}$$

Thus the geometry and the assumed displacement field are described in a similar fashion using the shape functions and nodal values.

4.3 Jacobian matrix

We next calculate the Jabobian Matrix **J** from the following expression

$$\mathbf{J} = \frac{\partial x}{\partial \xi} = \frac{\partial N_1}{\partial \xi} x_1 + \frac{\partial N_2}{\partial \xi} x_2 + \frac{\partial N_3}{\partial \xi} x_3. \tag{4.8}$$

This matrix will be required later in the formulation. In the present context, if the 2nd node is taken half way along the element, **J** is almost trivial and simply acts as a constant scaling factor, i.e.

$$\begin{aligned} \mathbf{J} &= (-\tfrac{1}{2} + \xi)x_1 - 2\xi x_2 + (\tfrac{1}{2} + \xi)x_3 \\ &= \frac{x_3 - x_1}{2} + \xi(x_1 + x_3 - 2x_2) \\ &= \frac{x_3 - x_1}{2} = \frac{L}{2} \end{aligned} \tag{4.9}$$

where L is the length of the element.

Later, in Chapters 5 and 6 when we examine 2D elements, we will use a more complicated Jacobian.

4.4 Strain definition

The strains are defined in terms of the nodal displacements and shape functions derivatives by the expression

$$\begin{bmatrix} \dfrac{\partial \theta}{\partial x} \\[2mm] \phi = -\dfrac{\partial w}{\partial x} + \theta \end{bmatrix} = \begin{bmatrix} 0 & \dfrac{\partial N_1}{\partial x} & 0 & \dfrac{\partial N_2}{\partial x} & 0 & \dfrac{\partial N_3}{\partial x} \\[3mm] -\dfrac{\partial N_1}{\partial x} & N_1 & -\dfrac{\partial N_2}{\partial x} & N_2 & -\dfrac{\partial N_3}{\partial x} & N_3 \end{bmatrix} \begin{bmatrix} w_1 \\ \theta_1 \\ w_2 \\ \theta_2 \\ w_3 \\ \theta_3 \end{bmatrix} \tag{4.10}$$

or

$$\boldsymbol{\varepsilon} = [\mathbf{B}_1, \mathbf{B}_2, \mathbf{B}_3]\, \delta^e = \mathbf{B}\delta^e$$

where $\partial\theta/\partial x$ is a pseudo-curvature and ϕ is the effective shear rotation.

The strain matrix $\mathbf{B}_i$ contains the shape function derivatives $\partial N_i/\partial x$ which may be calculated from the expression

$$\frac{\partial N_i}{\partial x} = \frac{\partial N_i}{\partial \xi} \cdot \frac{\partial \xi}{\partial x}$$

(4.11)

and $\partial \xi/\partial x$ may be obtained from the Jacobian matrix.

4.5 Stress/strain relationships

The "stress/strain" relationship for a beam of homogeneous isotropic material may be written as

$$\begin{bmatrix} M \\ Q \end{bmatrix} = \begin{bmatrix} EI & 0 \\ 0 & S \end{bmatrix} \begin{bmatrix} \dfrac{\partial \theta}{\partial x} \\ \phi \end{bmatrix}$$

(4.12)

or

$$\boldsymbol{\sigma} = \mathbf{D}\boldsymbol{\varepsilon}$$

where M is the bending moment and Q is the shear force.

4.6 Stiffness matrix evaluation

We now have all of the information necessary to calculate the element stiffness matrix $\mathbf{K}^e$ where, according to (1.28),

$$\mathbf{K}^e = \int [\mathbf{B}]^T \mathbf{D}\mathbf{B} \, \mathrm{d}x.$$

(4.13)

In fact, a typical submatrix of $\mathbf{K}^e$ linking nodes i and j may be evaluated from the expression

$$\mathbf{K}^e_{ij} = \int [\mathbf{B}_i]^T \mathbf{D}\mathbf{B}_j \det \mathbf{J} \, \mathrm{d}\xi$$

(4.14)

where

$$\mathrm{d}x = \det \mathbf{J} \, \mathrm{d}\xi.$$

This may be written as

$$\mathbf{K}_{ij} = \int \begin{bmatrix} 0 & -\dfrac{\partial N_i}{\partial x} \\ \dfrac{\partial N_i}{\partial x} & N_i \end{bmatrix} \begin{bmatrix} EI & 0 \\ 0 & S \end{bmatrix} \begin{bmatrix} 0 & \dfrac{\partial N_j}{\partial x} \\ -\dfrac{\partial N_j}{\partial x} & N_j \end{bmatrix} \det \mathbf{J} \, \mathrm{d}\xi,$$

(4.15)

or in the form

$$
\mathbf{K}_{ij} = \left[\begin{array}{c|c}
\int S\left(\dfrac{\partial N_i}{\partial x}\right)\left(\dfrac{\partial N_j}{\partial x}\right) \det \mathbf{J}\, d\xi & -\int SN_j\left(\dfrac{\partial N_i}{\partial x}\right) \det \mathbf{J}\, d\xi \\
\hline
-\int SN_i\left(\dfrac{\partial N_j}{\partial x}\right) \det \mathbf{J}\, d\xi & \begin{array}{c}\int EI\left(\dfrac{\partial N_i}{\partial x}\right)\left(\dfrac{\partial N_j}{\partial x}\right) \det \mathbf{J}\, d\xi \\ + \int SN_i N_j \det \mathbf{J}\, d\xi \end{array}
\end{array} \right] \quad (4.16)
$$

We could calculate these integrals explicitly. Instead, we use an efficient numerical integration technique known as Gaussian integration which is discussed in Section 4.9.

4.7 Consistent nodal forces

To represent a distributed lateral pressure load q/unit length in terms of nodal forces, we make use of the consistent nodal force formulae developed in Chapter 1 using Virtual Work. The applied nodal forces P_i and couples C_i may be represented by the vector

$$
\mathbf{F}^e = [P_1, C_1, P_2, C_2, P_3, C_3]^{\mathrm{T}}. \tag{4.17}
$$

For a distributed loading q,

$$
P_i = \int N_i q\, dx
$$

and

$$
C_i = 0. \tag{4.18}
$$

The integral for P_i may be calculated numerically using Gaussian integration.

It is interesting to note that this particular beam element does not require any nodal couples in the discrete representation of a distributed load.

4.8 Element stress resultants

When the nodal displacements have been calculated it is possible to calculate the stress resultants at the Gauss points of each element, using the expression

$$
\sigma(\xi_i) = \mathbf{D}(\xi_i)\, \mathbf{B}(\xi_i)\, \delta^e, \tag{4.19}
$$

where ξ_i are evaluated at the Gauss points. It is, of course, possible to calculate

σ at any point within the element and it is usual to calculate the stress resultants at the nodal points, i.e. $\xi_i = \xi_1, \xi_2$ or ξ_3. This calculation requires $\mathbf{D}$ and $\mathbf{B}$ to be evaluated at the nodal points. The sign convention for the stress resultants is given in Fig. 4.4.

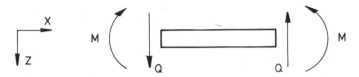

FIG. 4.4 Sign convention for bending moments and shear forces ($+$ VE sense indicated).

4.9 Numerical integration

We now digress slightly to consider the important topic of numerical integration.

In the finite element method, we must evaluate certain integrals such as the stiffness integrals $\int_A [\mathbf{B}]^T \mathbf{D} \mathbf{B} \, dA$ and the consistent load integral $\int_A Nq \, dA$.

In applications not covered in the book we might have to evaluate other integrals. For example, in structural dynamics the mass matrix would involve integrals of the type $\int_A [\mathbf{N}]^T \mathbf{N} \rho \, dA$, where ρ is the material mass density. In non-linear problems, quantities known as residual forces appear in integral form as $\int_A [\mathbf{B}]^T \sigma \, dA$.

In all such situations we choose numerical integration techniques because of their inherent advantages over analytical integration procedures. In particular, we choose Gauss–Legendre quadrature for its high accuracy and the ease with which it can be implemented.

It should be noted that an n-point rule integrates any polynomial of degree x^{2n-1}, or less, exactly.

In general, the one-dimensional Gaussian quadrature formula is written as

$$I_n = \int_{-1}^{+1} \phi(\xi) \, d\xi = \sum_{i=1}^{n} a_i \, \phi(\xi_i) \tag{4.20}$$

where a_i = weighting factor,
        ξ_i = coordinate of the ith integration point,
        n = total number of integration points.

A list of these is now given in tabular form below, where the Roman numerals refer to Gaussian sampling positions: e.g. I and II for a 2-point rule, I, II and III for a 3-point rule.

n	i	ξ_i	a_i
1	I	0	2
2	$\begin{cases} \text{I} \\ \text{II} \end{cases}$	$+1/\sqrt{3}$ $-1/\sqrt{3}$	$+1$ $+1$
3	$\begin{cases} \text{I} \\ \text{II} \\ \text{III} \end{cases}$	0 $+\sqrt{0.6}$ $-\sqrt{0.6}$	8/9 5/9 5/9
4	I	$\sqrt{\dfrac{3+\sqrt{4.8}}{7}}$	$\dfrac{1}{2}-\dfrac{\sqrt{30}}{36}$
	II	$-\sqrt{\dfrac{3+\sqrt{4.8}}{7}}$	$\dfrac{1}{2}-\dfrac{\sqrt{30}}{36}$
	III	$\sqrt{\dfrac{3-\sqrt{4.8}}{7}}$	$\dfrac{1}{2}+\dfrac{\sqrt{30}}{36}$
	IV	$-\sqrt{\dfrac{3-\sqrt{4.8}}{7}}$	$\dfrac{1}{2}+\dfrac{\sqrt{30}}{36}$

It should be noted that the sampling positions of the Gauss–Legendre rules coincide with the roots of the Legendre polynomials. Figure 4.5 shows the sampling position of a 2-point rule.

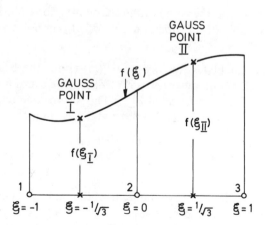

FIG. 4.5.   2-point Gaussian integration for the beam element.

In our program, subroutine GAUSSQ (described in Chapter 3) sets up the sampling point positions and weighting factors for numerical integration. N.B. One-point, 2-point and 3-point rules are allowed.

Using the 2-point rule, a typical integration in $\mathbf{K}_{ij}$ may be evaluated as

$$\int s\left(\frac{\partial N_i}{\partial x}\right)\left(\frac{\partial N_j}{\partial x}\right)\det \mathbf{J}\, d\xi =$$

$$s\left[a_{\mathrm{I}}\frac{\partial N_i}{\partial x}(\xi_{\mathrm{I}})\frac{\partial N_j}{\partial x}(\xi_{\mathrm{I}})\det \mathbf{J}(\xi_{\mathrm{I}}) + a_{\mathrm{II}}\frac{\partial N_i}{\partial x}(\xi_{\mathrm{II}})\frac{\partial N_j}{\partial x}(\xi_{\mathrm{II}})\det \mathbf{J}(\xi_{\mathrm{II}})\right]\qquad(4.21)$$

where $\partial N_i(\xi_{\mathrm{I}})/\partial x$ is the value of $\partial N_i/\partial x$ at $\xi = \xi_{\mathrm{I}}$ etc. This technique is used to calculate all the integrals in each of the $\mathbf{K}_{ij}$ submatrices and hence the complete element stiffness matrix $\mathbf{K}^e$ may be evaluated.

Using the 2-point rule, the consistent nodal forces may be calculated as

$$P_i = \int N_i q\, dx$$

$$= q(a_{\mathrm{I}}N_i(\xi_{\mathrm{I}})\det \mathbf{J}(\xi_{\mathrm{I}})$$

$$+ a_{\mathrm{II}}N_i(\xi_{\mathrm{II}})\det \mathbf{J}(\xi_{\mathrm{II}}).\qquad(4.22)$$

4.10 Subroutines

In this section, the following series of subroutines are presented:

SFR1 which calculates the value of the shape functions N_1, N_2 and N_3 and their derivatives $\partial N_1/\partial\xi$, $\partial N_2/\partial\xi$ and $\partial N_3/\partial\xi$ at any point within the beam element.

JACOB1 which calculates for any point (ξ) within the beam element the Cartesian shape function derivatives $\partial N_1/\partial x$, $\partial N_2/\partial x$ and $\partial N_3/\partial x$ using the Jacobian Matrix $\mathbf{J}$ and its inverse $[\mathbf{J}]^{-1}$. This subroutine requires the information calculated in SFR1 and also the nodal coordinates x_1, x_2 and x_3.

BMATB which calculates the strain matrix $\mathbf{B}$, at any point within the element using the shape functions N_1, N_2 and N_3 and their Cartesian derivatives $\partial N_1/\partial x$, $\partial N_2/\partial x$ and $\partial N_3/\partial x$.

MODB which calculates the components of the matrix of elastic rigidities $\mathbf{D}$ from the element material properties.

DBE which carries out the matrix multiplication $\mathbf{DB}$ and hence calculates the stress matrix at any point within the element.

STIFB which uses all of the above routines to calculate the stiffness and stress matrices for the beam element. It then stores this information on file for use at the solution stage of the program.

LOADB which uses SFR1 and JACOB1 to calculate the element consistent nodal forces and reads all load data.

STREB which calculates the stresses $\boldsymbol{\sigma}$ at the Gauss points after the displacements have been found using the expression $\boldsymbol{\sigma} = \mathbf{DB}\boldsymbol{\delta}^e$.

We described INPUT, NODEXY and GAUSSQ in Chapter 3, and we shall describe MASTER and FRONT later in Chapters 10 and 8 respectively. Here we concentrate on the evaluation of:

the element stiffness matrix and stress matrix,

the element consistent nodal forces,

the stress resultants using the nodal displacements.

Before describing each routine in detail, we list the main variables used in the evaluation of these three items.

Dictionary of variable names

ASDIS(MSVAB)	vector of nodal displacements calculated using FRONT—the sequence is $(w_1, \theta_1, w_2, \theta_2, w_3, \theta_3)$ etc.
BMATX(NSTRE,NEVAB)	the element strain matrix at any point ξ_P within the element $\mathbf{B} = [\mathbf{B}_1, \mathbf{B}_2, \mathbf{B}_3]$.
CARTD(NDIME,NNODE)	CARTesian shape function Derivatives associated with nodes of current element sampled at any point ξ_P within the element i.e. $$\left[\frac{\partial N_1}{\partial x}(\xi_P), \frac{\partial N_2}{\partial x}(\xi_P), \frac{\partial N_3}{\partial x}(\xi_P)\right].$$
COORD(MPOIN,NDIME)	COORDinates of nodal points.
DBMAT(NSTRE,NEVAB)	the result of the matrix multiplication $\mathbf{DB}$.
DERIV(NDIME,NNODE)	shape function DERIVatives associated with nodes of the current element—sampled at any point ξ_P within the element i.e. $$\left[\frac{\partial N_1}{\partial \xi}(\xi_P), \frac{\partial N_2}{\partial \xi}(\xi_P), \frac{\partial N_3}{\partial \xi}(\xi_P)\right].$$
DJACB	Determinant of the JACoBian matrix sampled at any point ξ_P within the element.

DLENG	an infinitesimal element of length at a Gauss point x the Gaussian weight coefficient.
DMATX(NSTRE,NSTRE)	the matrix of elastic rigidities **D**.
ELCOD(NDIME,NNODE)	local array of nodal Cartesian coordinates for the element currently under consideration.
ELDIS(NDOFN,NNODE)	nodal displacements associated with a particular element—to be used in stress resultant calculations.
ELOAD(MELEM,NEVAB)	nodal forces for each element.
ESTIF(NEVAB,NEVAB)	the Element STIFfness matrix $\mathbf{K}^e$.
EXISP	ξ coordinate of a sampling point— this is usually a Gauss point. N.B. EXISP = S in subroutine SFR1.
GPCOD(NDIME,NGASP)	local array of Cartesian coordinates of Gauss points for element currently under consideration.
IDIME,NDIME	Index, Number of DIMEnsions NDIME = 1 for a beam element.
IDOFN, NDOFN	Index, Number of Degrees Of Freedom per Node. NDOFN = 2 for a beam element.
IELEM,MELEM*,NELEM	Index, Maximum, Number of ELEMents. MELEM never actually appears in the program and is used here simply for convenience. MELEM = 25 for the beam program.
IEVAB,JEVAB,NEVAB	Index, Index, Number of variables per element. NEVAB = 6 for a beam element.
IGAUS,JGAUS,MGAUS,NGAUS	Index, Index, Maximum, Number of GAUSs rule adopted. MGAUS never actually appears in the program and is used here simply for convenience. MGAUS = NGAUS = 2 for the beam program.
INODE,NNODE	Index, Number of NODes per Element NNODE = 3 for a beam element.

ISTRE,NSTRE	Index, Number of STREss resultant types. MSTRE = 2 for a beam element (i.e. M and Q).
KGASP,NGASP	Kounter, Number of GAuSs Points used. NGASP = 2 for a beam element.
LNODE	node currently under consideration.
LNODS(MELEM,NNODE)	eLement NODe numberS listed for each element
LPROP	material set of element currently under consideration.
MATNO(MELEM)	Material set Numbers for each element.
MMATS,NMATS	Maximum, Number of MATerial Sets MMATS is never actually used in the program and is given here simply for convenience. MMATS = 10 for the beam program.
MPOIN	Maximum number of nodal POINts. MPOIN is never actually used in the program and is given here simply for convenience. MPOIN = 50 for the beam program.
MSVAB	Maximum number of Structural VAriaBles MSVAB = MPOIN × NDOFN. MSVAB is never actually used in the program and is given here simply for convenience. MSVAB = 100 for the beam program.
POSGP(MGAUS)	ξ coordinates of Gauss points.
PROPS(MMATS,NPROP)	Material PROPertieS for each material set.
SHAPE(NNODE)	SHAPE functions associated with each node of the current element sampled at any point ξ_P within the element i.e. $[N_1(\xi_P), N_2(\xi_P), N_3(\xi_P)]^T$.
SMATX(NSTRE,NEVAB,MGASP)	contains DBMAT for each Gauss point–element stress matrix.
STRSG(NSTRE)	STReSs resultants at Gauss point ξ_P for current element i.e. $[M(\xi_P), Q(\xi_P)]^T$.
WEIGP(MGAUS)	WEIghting factors for Gauss Points.

Variables which are underlined merely denote maximum values following the convention of Chapter 1.

4.10.1 Subroutine STIFB

A schematic representation of the most important routine STIFB is now presented. The information in brackets is transferred between the two subroutines.

```
        SUBROUTINE STIFB
        DIMENSIONS AND COMMON BLOCKS
→       ENTER LOOP COVERING ALL ELEMENTS
        RETRIEVE ELEMENT GEOMETRY
            AND MATERIAL PROPERTIES
        ZERO STIFFNESS ARRAY
        CALL MODB (ELEMENT MATERIAL PROPERTIES)—This
            returns D the matrix of elastic rigidities
→       ENTER LOOP COVERING ALL INTEGRATION POINTS
        LOOK UP COORDINATES OF POINTS (ξ)
        CALL SFR1 (ξ)—This returns shape function with ξ derivatives
        CALL JACOB1 (ξ, GEOMETRY AND SHAPE FUNCTIONS
            WITH x DERIVATIVES)—This returns Jacobian J and
            Cartesian shape function derivatives
        CALL BMATB (CARTESIAN SHAPE FUNCTION DERIVA-
            TIVES)—This returns B the strain matrix
        CALL DBE (D and B)—This returns DB
        COMPUTE[B]ᵀ DB det J × INTEGRATING FACTOR AND
            ASSEMBLE INTO ELEMENT STIFFNESS ARRAY
        ASSEMBLE DB INTO STRESS MATRIX ARRAY
        WRITE STIFFNESS AND STRESS MATRICES ONTO FILE
            FOR USE LATER IN SOLUTION ROUTINES (It is auto-
            matically assumed that backing store facilities are available,
            and frequent use is made of disc files throughout the programs)
        RETURN
        END
```

We now repeat in FORTRAN what we have illustrated schematically above.

Input/Output diagram

```
                    INPUT                         OUTPUT
                 ┌ NELEM         → STIFB →   ESTIF(NEVAB,NEVAB)]  F(1)
                 │ NDIME
                 │ NNODE                     SMATX(NSTRE,NEVAB,NGASP) ⎤ F(3)
   C(CONTRO)     │ NEVAB                     GPCOD(NDIME,NGASP)       ⎦
                 │ NGAUS
                 └ NSTRE

                 ┌ PROPS(NMATS,NPROP)
                 │ MATNO(NELEM)
   C(LGDATA)     │ LNODS(NELEM,NNODE)
                 │ COORD(NPOIN,NDIME)
                 │ WEIGP(NGAUS)
                 └ POSGP(NGAUS)

   C(WORK)       ┌ BMATX(NSTRE,NEVAB)
                 └ DBMAT(NSTRE,NEVAB)

   A(JACOB1)     ⎡ DJACB
```

Annotated FORTRAN listing

```
              SUBROUTINE STIFB
              DIMENSION ESTIF(6,6)

        ┌─────────────────────────┐
        │   COMMON  BLOCKS        │
        └─────────────────────────┘

    C
    C*** LOOP OVER EACH ELEMENT
    C
  ┌─────── DO 70 IELEM=1,NELEM                          1*
  │        LPROP=MATNO(IELEM)                           2*
  │  C
  │  C*** EVALUATE THE COORDINATES OF THE ELEMENT
  │  C    NODAL POINTS
  │  C
  │  ┌───── DO 10 INODE=1,NNODE                    ⎫
  │  │      LNODE=LNODS(IELEM,INODE)               ⎪
  │  │┌──── DO 10 IDIME=1,NDIME                     ⎬ 3*
  │  ││     ELCOD(IDIME,INODE)=COORD(LNODE,IDIME)  ⎪
  │  └┴ 10 CONTINUE                                 ⎭
  │  C
  │  C*** INITIALIZE THE ELEMENT STIFFNESS MATRIX
  │  C
  │  ┌───── DO 20 IEVAB=1,NEVAB                     ⎫
  │  │┌──── DO 20 JEVAB=1,NEVAB                     ⎬ 4*
  │  ││     ESTIF(IEVAB,JEVAB)=0.0                  ⎪
  │  └┴ 20 CONTINUE                                 ⎭
  │  C
  │  C*** EVALUATE THE D-MATRIX
  │  C
  │        CALL MODB(LPROP)                           5*
  │        KGASP=0                                     6*
```

```
C
C*** ENTER LOOPS FOR NUMERICAL INTEGRATION
C
       DO 50 IGAUS=1,NGAUS                        7*
       KGASP=KGASP+1                              8*
       EXISP=POSGP(IGAUS)                         9*
C
C*** EVALUATE THE SHAPE FUNCTIONS,
C    ELEMENTAL LENGTHS,ETC.
C
       CALL SFR1(EXISP)                           10*
       CALL JACOB1(IELEM,DJACB,KGASP)             11*
       DLENG=DJACB*WEIGP(IGAUS)                   12*
C
C*** EVALUATE THE B AND DB MATRICES
C
       CALL BMATB                                 13*
       CALL DBE                                   14*
C
C*** CALCULATE THE ELEMENT STIFFNESSES
C
       DO 30 IEVAB=1,NEVAB
       DO 30 JEVAB=IEVAB,NEVAB
       DO 30 ISTRE=1,NSTRE
       ESTIF(IEVAB,JEVAB)=ESTIF(IEVAB,JEVAB)+     15*
      .BMATX(ISTRE,IEVAB)*DBMAT(ISTRE,JEVAB)*
      .DLENG
    30 CONTINUE
C
C*** STORE THE COMPONENTS OF THE DB MATRIX
C    FOR THE ELEMENT
C
       DO 40 ISTRE=1,NSTRE
       DO 40 IEVAB=1,NEVAB
       SMATX(ISTRE,IEVAB,KGASP)=                  16*
      .DBMAT(ISTRE,IEVAB)
    40 CONTINUE
    50 CONTINUE                                   17*
C
C*** CONSTRUCT THE LOWER TRIANGLE OF THE
C    STIFFNESS MATRIX
C
       DO 60 IEVAB=1,NEVAB                         18*
       DO 60 JEVAB=1,NEVAB
       ESTIF(JEVAB,IEVAB)=ESTIF(IEVAB,JEVAB)
    60 CONTINUE
C
C*** STORE THE STIFFNESS MATRIX,STRESS MATRIX
C    AND SAMPLING POINT COORDINATES
C    FOR EACH ELEMENT ON DISC FILE
C
       WRITE(1) ESTIF                             19*
       WRITE(3) SMATX,GPCOD                       20*
    70 CONTINUE                                   21*
       RETURN
       END
```

1* Loop over each element.
2* Determine material property set for current element.
3* Create local array ELCOD of nodal coordinates for current element.
4* Zero stiffness array ESTIF.
5* Call MODB to calculate elastic rigidities of beam element—note that argument LPROP is the material property set number for current element.
6* Initialise Gauss point counter.
7* Loop over each Gauss point.
8* Increment Gauss point counter.
9* Determine ξ coordinate of current Gauss point.
10* Call SFR1 to obtain shape functions and their derivatives for current Gauss point—note argument EXISP is ξ coordinate of current Gauss point.
11* Call JACOB1 to obtain Cartesian shape function derivatives and determinant of Jacobian matrix at current Gauss point—note argument is current element number, determinant of Jacobian matrix and number of Gauss point.
12* Calculate an element of length multiplied by Gauss weight.
13* Call BMATB to calculate strain matrix **B** at current Gauss point.
14* Call DBE to multiply **D** by **B**.
15* Calculate the upper triangle of the element stiffness matrix ESTIF $= \int [\mathbf{B}]^{\mathrm{T}} \, \mathbf{DB} \, \mathrm{d}x$ adding the contribution from each Gauss point.
16* Store stress matrix for each Gauss point in SMATX.
17* End Gauss loop.
18* Construct lower triangle of stiffness matrix.
19* Write stiffness matrix ESTIF on file 1.
20* Write stress matrix SMATX and Gauss point coordinates on file 3.
21* End element loop.

4.10.2 Subroutine SFR1

A straightforward subroutine which calculates the values of the shape functions N_1, N_2 and N_3 and their derivatives $\partial N_1/\partial\xi$, $\partial N_2/\partial\xi$ and $\partial N_3/\partial\xi$ at any sampling point within the element. This sampling point is usually the Gauss point.

Input/Output diagram

$$
A \begin{bmatrix} \begin{matrix} \text{EXISP} \\ = \text{S} \end{matrix} & \rightarrow & \boxed{\text{SFR1}} & \rightarrow & \begin{matrix} \text{SHAPE(NNODE)} \\ \text{DERIV(NDIME,NNODE)} \end{matrix} \end{bmatrix} \text{C(WORK)}
$$

Annotated FORTRAN listing

```
            SUBROUTINE SFR1(S)
      C
      C*** CALCULATES SHAPE FUNCTIONS AND THEIR
      C    DERIVATIVES FOR 1D ELEMENTS
      C
```

┌─────────────────────┐
│ COMMON BLOCKS   │
└─────────────────────┘

```
      S2=S*2.0                                   } 1 *
      SS=S*S
      C
      C*** SHAPE FUNCTIONS
      C
      SHAPE(1)=(-S+SS)/2.0
      SHAPE(2)=1.0-SS                            } 2 *
      SHAPE(3)=(S+SS)/2.0
      C
      C*** SHAPE FUNCTION DERIVATIVES
      C
      DERIV(1,1)=(-1.0+S2)/2.0
      DERIV(1,2)=-S2                             } 3 *
      DERIV(1,3)=(1.0+S2)/2.0
      RETURN
      END
```

1* Evaluate 2ξ and ξ^2 at sampling point.
2* Evaluate shape functions N_1, N_2 and N_3 at sampling point.
3* Evaluate shape function derivatives $\partial N_1/\partial\xi$, $\partial N_2/\partial\xi$ and $\partial N_3/\partial\xi$ at sampling point.

4.10.3 Subroutine JACOB1

A straightforward subroutine which calculates the Cartesian shape function derivatives, the Jacobian matrix and the coordinates of the sampling point (usually a Gauss point).

Input/Output diagram

	INPUT		OUTPUT	
C(CONTRO)	NDIME NNODE	→ JACOB1 →	DJACB]	
C(LGDATA)	SHAPE(NNODE) DERIV(NDIME,NNODE)		GPCOD(NDIME,NGASP) CARTD(NDIME,NNODE)	C(WORK)
A	IELEM KGASP			

Annotated FORTRAN listing

```
      SUBROUTINE JACOB1(IELEM,DJACB,KGASP)
C
C*** CALCULATES COORDINATES OF GAUSS POINTS
C    AND THE JACOBIAN MATRIX AND ITS DETERMINANT
C    AND THE INVERSE FOR 1D ELEMENTS
C
```

┌─────────────────────┐
│ COMMON BLOCKS   │
└─────────────────────┘

```
      DJACB=0.0                                              1*
      DO 10 IDIME=1,NDIME
      GPCOD(IDIME,KGASP)=0.0                                 2*
   10 CONTINUE
C
C*** CALCULATES COORDINATES OF SAMPLING POINT
C
      DO 20 IDIME=1,NDIME
      DO 20 INODE=1,NNODE
      GPCOD(IDIME,KGASP)=GPCOD(IDIME,KGASP)+                 3*
     .SHAPE(INODE)*ELCOD(IDIME,INODE)
   20 CONTINUE
C
C*** CALCULATE DETERMINANT OF JACOBIAN MATRIX
C
      DO 30 INODE=1,NNODE
      DJACB=DJACB+DERIV(1,INODE)*ELCOD(1,INODE)              4*
   30 CONTINUE
      IF(DJACB)40,40,50
   40 WRITE(6,900) IELEM                                     5*
      STOP
C
C*** CALCULATE CARTESIAN DERIVATIVES
C
   50 DO 60 INODE=1,NNODE
      CARTD(1,INODE)=DERIV(1,INODE)/DJACB                    6*
   60 CONTINUE
  900 FORMAT(//,10X,
     .36H PROGRAM HALTED IN SUBROUTINE JACOB1,
     ./,11X,22H ZERO OR NEGATIVE AREA,/,10X,
     .16H ELEMENT NUMBER ,I5)
      RETURN
      END
```

1* Set DJACB equal to zero.
2* Set GPCOD equal to zero.
3* Evaluate coordinates of sampling point using the shape functions and
    nodal coordinates.
4* Evaluate the determinant of the Jacobian matrix DJACB.
5* If DJACB is not positive write a message and stop—otherwise continue.
6* Evaluate the Cartesian shape function derivatives at sampling point.

4.10.4 Subroutine BMATB

Subroutine BMATB calculates the strain matrix **B** at the sampling point
using the value of the shape functions N_1, N_2 and N_3 and their Cartesian
derivatives previously calculated at the sampling point.

Input/Output diagram

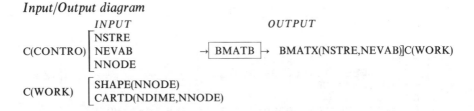

Annotated FORTRAN listing

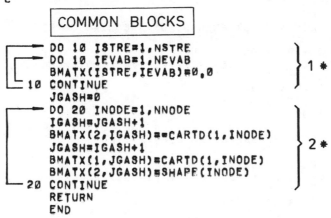

1* Zero the strain matrix **B**.

2* Construct the non-zero components of **B** noting that

$$\mathbf{B} = \begin{bmatrix} 0 & \dfrac{\partial N_1}{\partial x} & 0 & \dfrac{\partial N_2}{\partial x} & 0 & \dfrac{\partial N_3}{\partial x} \\ -\dfrac{\partial N_1}{\partial x} & N_1 & -\dfrac{\partial N_2}{\partial x} & N_2 & -\dfrac{\partial N_3}{\partial x} & N_3 \end{bmatrix}.$$

4.10.5 Subroutine MODB

Subroutine MODB calculates the coefficients of the matrix of elastic rigidities
D for the current element using the element material properties.

Input/Output diagram

```
              INPUT                           OUTPUT
C(CONTRO) [NSTRE          →| MODB |→  DMATX(NSTRE,NSTRE)] C(WORK)

C(LGDATA)[PROPS(MMATS,NPROP)

A          [LPROP
```

Annotated FORTRAN listing

```
        SUBROUTINE MODB(LPROP)
  C
  C*** CALCULATES D-MATRIX FOR BEAM ELEMENT
  C
```

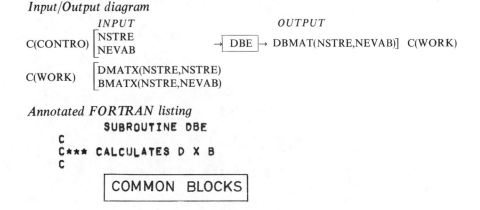

```
       ┌───► DO 10 ISTRE=1,NSTRE         ⎫
       │ ──► DO 10 JSTRE=1,NSTRE         ⎬ 1*
       │     DMATX(ISTRE,JSTRE)=0.0      ⎭
       └─10 CONTINUE
             DMATX(1,1)=PROPS(LPROP,1)   ⎫
             DMATX(2,2)=PROPS(LPROP,2)   ⎬ 2*
             RETURN                      ⎭
             END
```

1* Zero the matrix of elastic rigidities **D**.
2* Construct non-zero components of **D**
   i.e. $D(1, 1) = EI$ flexural rigidity
   $D(2, 2) = S$ shear rigidity.

4.10.6 Subroutine DBE

This subroutine simply multiplies **D** by **B**.

Input/Output diagram

```
              INPUT                           OUTPUT
          ⎡NSTRE
C(CONTRO) ⎢                    →| DBE |→ DBMAT(NSTRE,NEVAB)]  C(WORK)
          ⎣NEVAB

          ⎡DMATX(NSTRE,NSTRE)
C(WORK)   ⎢
          ⎣BMATX(NSTRE,NEVAB)
```

Annotated FORTRAN listing

```
        SUBROUTINE DBE
  C
  C*** CALCULATES D X B
  C
        COMMON  BLOCKS
```

```
         DO 10 ISTRE=1,NSTRE
         DO 10 IEVAB=1,NEVAB
         DBMAT(ISTRE,IEVAB)=0.0
         DO 10 JSTRE=1,NSTRE
         DBMAT(ISTRE,IEVAB)=DBMAT(ISTRE,IEVAB)+    1*
       .DMATX(ISTRE,JSTRE)*BMATX(JSTRE,IEVAB)
      10 CONTINUE
         RETURN
         END
```

1* Evaluate **D** multiplied by **B**.

4.10.7 Subroutine LOADB

This subroutine calculates the consistent nodal forces for a uniformly distributed load over a beam element. At this stage, we do not deal with other types of loads, e.g. point loads, gravity loads, etc. These will be dealt with for 2D elements only in Chapter 7.

Input/Output diagram

```
              INPUT                    OUTPUT
            ┌ NELEM        → LOADB → ELOAD(NELEM,NEVAB)]C(LGDATA)
            │ NNODE
            │ NDIME
  C(CONTRO) │ NGAUS
            │ NEVAB
            └ NDOFN

            ┌ MATNO(NELEM)
            │ LNODS(NELEM,NNODE)
  C(LGDATA) │ COORD(NPOIN-NDIME)
            └ PROPS(NMATS,NPROP)
```

Annotated FORTRAN listing

```
          SUBROUTINE LOADB
      C
      C*** CALCULATE NODAL FORCES FOR BEAM ELEMENT
      C
```

┌─────────────────────┐
│   COMMON BLOCKS   │
└─────────────────────┘

```
      C
      C*** LOOP OVER EACH ELEMENT
      C
         DO 50 IELEM=1,NELEM                          1*
         LPROP=MATNO(IELEM)                           2*
      C
      C*** EVALUATE THE COORDINATES OF THE ELEMENT
      C    NODAL POINTS
      C
```

```
        ┌──►DO 10 INODE=1,NNODE              ┐
        │   LNODE=LNODS(IELEM,INODE)         │
      ┌─┼──►DO 10 IDIME=1,NDIME              ├ 3*
      │ │   ELCOD(IDIME,INODE)=COORD(LNODE,IDIME) │
      └─┼─10 CONTINUE                        ┘
        │   UDLOD=PROPS(LPROP,3)                   4*
        │   IF(UDLOD.EQ.0.0) GO TO 50              5*
      ┌─┼──►DO 20 IEVAB=1,NEVAB             ┐
      │ │   ELOAD(IELEM,IEVAB)=0.0          ├ 6*
      └─┼─20 CONTINUE                        ┘
        │   KGASP=0                                7*
      C
      C*** ENTER LOOPS FOR NUMERICAL INTEGRATION
      C
      ┌─┼──►DO 40 IGAUS=1,NGAUS                     8*
      │ │   KGASP=KGASP+1                           9*
      │ │   EXISP=POSGP(IGAUS)                     10*
      │ C
      │ C*** EVALUATE THE SHAPE FUNCTIONS AT THE
      │ C    SAMPLING POINTS AND ELEMENTAL LENGTH
      │ C
      │ │   CALL SFR1(EXISP)                       11*
      │ │   CALL JACOB1(IELEM,DJACB,KGASP)         12*
      │ │   DLENG=DJACB*WEIGP(IGAUS)               13*
      │ C
      │ C*** CALCULATE LOADS AND ASSOCIATE WITH
      │ C    ELEMENT NODAL POINTS
      │ C
      │ ┌─►DO 30 INODE=1,NNODE             ┐
      │ │  NPOSN=(INODE-1)*NDOFN+1          │
      │ │  ELOAD(IELEM,NPOSN)=ELOAD(IELEM,NPOSN)+ ├ 14*
      │ │  .SHAPE(INODE)*UDLOD*DLENG        │
      │ └─30 CONTINUE                        ┘
      └───40 CONTINUE                             15*
      └──── 50 CONTINUE
      C
      C*** WRITE ELEMENT NODAL FORCES
      C
          WRITE(6,900)
      900 FORMAT(1H0,5X,
         .36H TOTAL NODAL FORCES FOR EACH ELEMENT)
      ┌──►DO 60 IELEM=1,NELEM
      │   WRITE(6,905) IELEM,
      │   .(ELOAD(IELEM,IEVAB),IEVAB=1,NEVAB)
      └─60 CONTINUE
      905 FORMAT(1X,I4,5X,3F12.4/(10X,3F12.4))
          RETURN
          END
```

1* Loop over each element.
2* Determine material property set for the current element.
3* Create local array ELCOD of nodal coordinates for current element.
4* Determine the value of the distributed load over the current element.

5* If this load is zero skip to next element.
6* Zero space for loads.
7* Set Gauss point counter to zero.
8* Loop over each Gauss point.
9* Increment Gauss point counter.
10* Determine ξ coordinate of current Gauss point.
11* Call SFR1 to obtain shape functions and their derivatives for current Gauss point.
12* Call JACOB1 to obtain Cartesian shape function derivatives and determinant of Jacobian matrix for current Gauss point.
13* Calculate an element of length multiplied by Gauss weight.
14* For each node calculate consistent nodal force due to uniformly distributed load.
15* End element loop.

4.10.8 Subroutine STREB

Subroutine STREB evaluates the stresses at the Gauss points once the nodal displacements are known, i.e. after the solution.

Input/Output diagram

```
                    INPUT                              OUTPUT
                 ┌ NELEM ┐        →│ STREB │→      STRSG(NSTRE)      ┐
                 │ NNODE │                          GPCOD(NDIME,NGASP) │ F(6)
   C(CONTRO) │ NDOFN │                                             ┘
                 │ NGAUS │
                 │ NSTRE │
                 └ NDIME ┘

   C(LGDATA) [ASDIS(MPOIN × NDOFN)

   F(3)          ┌ SMATX(NSTRE,NEVAB,NGASP)
                 └ GPCOD(NDIME,NGASP)
```

Annotated FORTRAN listing

```
        SUBROUTINE STREB
C
C*** CALCULATE THE STRESS RESULTANTS AT
C   THE GAUSS POINTS FOR BEAM
C
        DIMENSION ELDIS(2,3),STRSG(2)
```

┌─────────────────────┐
│ COMMON BLOCKS   │
└─────────────────────┘

```
        WRITE(6,900)
        WRITE(6,905)                              1 *
```

```
      C
      C*** LOOP OVER EACH ELEMENT
      C
            DO 40 IELEM=1,NELEM                                2*
      C
      C*** READ THE STRESS MATRIX , SAMPLING POINT
      C    COORDINATES FOR THE ELEMENT
      C
            READ(3) SMATX,GPCOD                                3*
            WRITE(6,910) IELEM                                 4*
      C
      C*** IDENTIFY THE DISPLACEMENTS OF THE
      C    ELEMENT NODAL POINTS
      C
            DO 10 INODE=1,NNODE
            LNODE=LNODS(IELEM,INODE)
            NPOSN=(LNODE-1)*NDOFN                          }  5*
            DO 10 IDOFN=1,NDOFN
            NPOSN=NPOSN+1
            ELDIS(IDOFN,INODE)=ASDIS(NPOSN)
         10 CONTINUE
            KGASP=0                                            6*
      C
      C*** ENTER LOOPS OVER EACH SAMPLING POINT
      C
            DO 30 IGAUS=1,NGAUS                                7*
            KGASP=KGASP+1                                      8*
            DO 20 ISTRE=1,NSTRE
            STRSG(ISTRE)=0.0
            KGASH=0
      C
      C*** COMPUTE THE STRESS RESULTANTS
      C
            DO 20 INODE=1,NNODE
            DO 20 IDOFN=1,NDOFN
            KGASH=KGASH+1                                  }  9*
            STRSG(ISTRE)=STRSG(ISTRE)+
           .SMATX(ISTRE,KGASH,KGASP)
           .*ELDIS(IDOFN,INODE)
         20 CONTINUE
      C
      C*** OUTPUT THE STRESS RESULTANTS
      C
            WRITE(6,915) KGASP,
           .(GPCOD(IDIME,KGASP),IDIME=1,NDIME)            } 10*
           .,(STRSG(ISTRE),ISTRE=1,NSTRE)
         30 CONTINUE
         40 CONTINUE                                          11*
        900 FORMAT(/,10X,8HSTRESSES,/)
        905 FORMAT(1H0,4HG.P.,2X,7HX=COORD,3X,
           .8HX=MOMENT,4X,10HXZ=S.FORCE)
        910 FORMAT(/,5X,12HELEMENT NO.=,I5)
        915 FORMAT(I5,F10.4,2E12.5)
            RETURN
            END
```

  1* Write title.
  2* Loop over each element.
  3* Retrieve stress matrix and Gauss point coordinates from file.
  4* Write element number.
  5* Assemble element displacements into local array ELDIS from global array ASDIS.
  6* Initialise Gauss point counter.
  7* Loop over each Gauss point.
  8* Increment Gauss point counter.
  9* Calculate stress resultants $\boldsymbol{\sigma} = \mathbf{DB}\boldsymbol{\delta}^e$.
 10* Write stress resultants and Gauss point coordinates for each Gauss point.
 11* End element loop.

5

Element Characteristics for Plane Stress/Strain

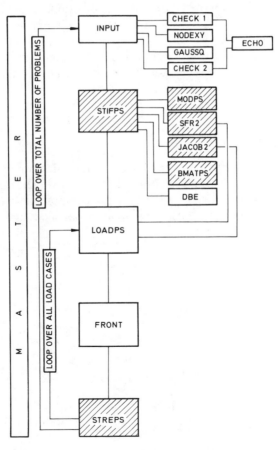

5.1 Introduction and basic theory

In this chapter, we introduce a two-dimensional parabolic isoparametric element which can be employed in the solution of problems satisfying either the conditions of plane stress of those of plane strain. Often in stress analysis, a full three dimensional treatment of a problem can be avoided by assuming the structure to be adequately represented if plane stress or plane strain conditions are adopted. Historically, plane solutions were the first attempted by the finite element method with solution being performed using simplex elements [1]. Such solutions illustrate all the essential features of the finite element process without introducing the complications encountered in other applications, such as shell analysis. For this reason the solution of plane problems by the use of parabolic isoparametric elements is the first two-dimensional application considered in this text.

The basic definitions and applicable expressions can be found in any standard text on the theory of elasticity (e.g. [2]) and only a brief outline of the assumed conditions is included here. A fully three-dimensional situation reduces to a two-dimensional problem if all quantities are independent of one of the coordinate directions, usually assumed be the z axis. Furthermore, for a planar condition to exist, all body forces and surface forces acting on the solid must act in the xy plane (i.e. have no z component).

5.1.1 Plane strain

If, in addition to the above assumptions, the normal strain in the z direction, ε_z, is zero then a plane strain condition is said to exist. In this case the only non-zero stress components are the in-plane components σ_x, σ_y, τ_{xy} and the through-thickness stress σ_z. All equations are satisfied and this is an exact

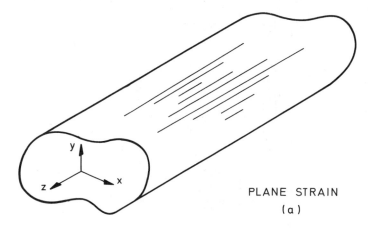

PLANE STRAIN

(a)

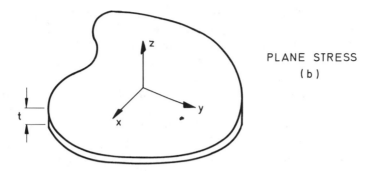

FIG. 5.1.   Prismatic solids satisfying plane conditions.

theory within the framework of three-dimensional elasticity.

Practically, plane strain conditions are achieved if:

The thickness of the solid (in the z direction) is large in comparison with the representative x and y dimension (Fig. 5.1(a)). Even if the ends of the cylinder are postulated to be free from traction, solution can still be achieved by superimposing a uniform axial stress field and pure bending distribution which statically balances the axial stress field, σ_z, obtained by assuming ε_z to be zero. (Any local end difference is accounted for by St. Venant's principle.)

Or, the ends of the cylindrical solid are constrained from moving in the z direction, e.g. a gravity dam or shafts with end thrust bearings.

5.1.2 Plane stress

On the other hand, if it is assumed that the normal stress in the z direction, σ_z, is zero, a plane stress condition exists. In this case the in-plane stress components σ_x, σ_y, τ_{xy} are again non-zero together with the normal strain component, ε_z. This is an approximate theory since the limitations placed on the variation of the ε_z strain over the xy plane cannot generally be satisfied and τ_{xz} and τ_{yz} are non-zero. However it is assumed that they can be neglected for analysis.

Practically this condition can be achieved if:

The body is thin (i.e. it is small in comparison with the representative x and y dimensions as shown in Fig. 5.1(b)).

Also there are no surface forces acting on the end faces $z = \pm t/2$.

Thus in both cases, the only nodal displacement components are those in the x and y directions, denoted by u and v respectively. Both variables are independent of the z direction, as indeed are all other quantities.

In the description of the element matrices, etc., we shall follow the mode of presentation employed in the previous chapter for the beam element. Loads, however, will be dealt with separately in Chapter 7 since a detailed explanation is required. By the end of this chapter we shall have sufficient information to enable us to calculate the element stiffness and stress matrices for plane stress/strain applications. Together with subroutines essential for calculating the element stiffness matrices, we will also develop a subroutine which will evaluate the stresses as soon as the nodal displacements have been found.

5.2 Shape functions

The initial step of any finite element analysis is the unique description of the unknown function δ (in our case the displacement field) within each element in terms of n parameters δ_i associated generally with the values of this function at nodal points in the form first stated in Chapter 1

$$\delta = \sum_{i=1}^{n} N_i \delta_i \tag{5.1}$$

in which N_i depend on the spatial coordinates and are known collectively as the shape functions matrices.

With the displacements known at all points within the element, the strains at any point can be determined by the relationship

$$\varepsilon = \sum_{i=1}^{n} B_i \delta_i \tag{5.2}$$

where the strain matrix B_i is generally composed of derivatives of the shape functions.

The efficiency of any particular element type used will depend on how well the shape functions are capable of representing the true displacement field. The choice of appropriate shape functions is however not arbitrary and there are two minimum conditions which must be satisfied in order to ensure convergence of the solution to the correct result as the finite element mesh is refined:

Shape functions must guarantee continuity of the function between elements (known as the *continuity* condition).

In the limit as the element size is reduced to infinitesimal dimensions, the shape function must be able to reproduce a constant strain condition through the element. Thus the unknown function must be able to take up in the limit any linear form throughout the element (known as the *constant strain* condition).

The isoparametric family are a group of elements in which the shape functions are used to define the geometry as well as the displacement field.

For plane stress or plane strain applications, the displacement fields $u(\xi, \eta)$ and $v(\xi, \eta)$ throughout the element are defined using two displacement degrees of freedom u_i and v_i at each of the eight nodes and a quadratic interpolation scheme.

We make use of the natural coordinate system (ξ, η) which allows us to use elements with curvilinear shapes. Using a similar procedure to that developed in Chapter 4 we define the coordinate values $x(\xi, \eta)$ and $y(\xi, \eta)$ at any point (ξ, η) within the element by the expressions

$$x(\xi, \eta) = \sum_{i=1}^{8} N_i(\xi, \eta) \cdot x_i$$

and   (5.3)

$$y(\xi, \eta) = \sum_{i=1}^{8} N_i(\xi, \eta) \cdot y_i$$

where (x_i, y_i) are the coordinates of node i, and where the two dimensional quadratic shape functions are given as

$$N_i(\xi, \eta) = -\tfrac{1}{4}(1 - \xi)(1 - \eta)(1 + \xi + \eta)$$
$$N_2(\xi, \eta) = \tfrac{1}{2}(1 - \xi^2)(1 - \eta)$$
$$N_3(\xi, \eta) = \tfrac{1}{4}(1 + \xi)(1 - \eta)(\xi - \eta - 1)$$
$$N_4(\xi, \eta) = \tfrac{1}{2}(1 + \xi)(1 - \eta^2)$$
$$N_5(\xi, \eta) = \tfrac{1}{4}(1 + \xi)(1 + \eta)(\xi + \eta - 1)$$
$$N_6(\xi, \eta) = \tfrac{1}{2}(1 - \xi^2)(1 + \eta)$$
$$N_7(\xi, \eta) = \tfrac{1}{4}(1 - \xi)(1 + \eta)(-\xi + \eta - 1)$$
$$N_8(\xi, \eta) = \tfrac{1}{2}(1 - \xi)(1 - \eta^2) \tag{5.4}$$

the nodal numbering being as shown in Fig. 5.2. (points indicated $\bigcirc$).

As with the one dimensional shape functions, each of these shape functions has a value of unity at the node to which it is related. They also have the property that their sum at any point within an element is also equal to unity, since it is required that a rigid body displacement of the element results in no element straining. There are also some other important shape function properties which will be discussed later.

At this point, we must pause to give some attention to the orientation of the

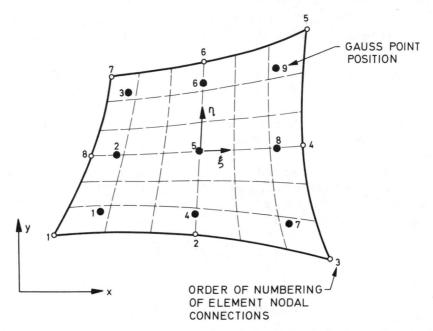

FIG. 5.2.   Orientation of local axes ξ, η, and order of Gauss point numbering for two dimensional parabolic isoparametric elements.

local coordinate directions ξ and η since the shape functions defined in (5.4) are clearly dependent on the orientation of ξ and η. The ξ, η variables are curvilinear coordinates and as such their direction will vary with position. However their general directions are always known relative to the element sides. The expressions quoted in (5.4) are based upon the following dependence between the ordering of the element nodal connection numbers and the local axes, ξ, η.

The nodes of an element are input in an anticlockwise sequence starting from any corner node, as previously mentioned in Chapter 3.

The positive ξ axis is then in the direction defined by moving along an element edge from the 1st element nodal connection number, through the 2nd to the 3rd.

The positive η axis is in the direction of the element edge from the 3rd nodal connection number, through the 4th to the 5th number.

Thus, with regard to Fig. 5.2, when the element topology is defined in the sequence indicated by nodal numbers 1 to 8, the directions of the ξ and η axes are as shown.

To calculate the displacements $u(\xi, \eta)$ and $v(\xi, \eta)$ at any point within the element, we make use of the expressions

$$u(\xi, \eta) = \sum_{i=1}^{8} N_i(\xi, \eta) \cdot u_i$$

$$v(\xi, \eta) = \sum_{i=1}^{8} N_i(\xi, \eta) \cdot v_i. \tag{5.5}$$

As with the one dimensional case we have used the same method to describe both the element displacement field and the element geometry, which gives rise to the term *isoparametric* element. The derivatives of a function $f(\xi, \eta)$ with respect to ξ and η can be obtained as follows

$$\frac{\partial f}{\partial \xi}(\xi, \eta) = \sum_{i=1}^{8} \frac{\partial N_i}{\partial \xi} \cdot f_i$$

$$\frac{\partial f}{\partial \eta}(\xi, \eta) = \sum_{i=1}^{8} \frac{\partial N_i}{\partial \eta} \cdot f_i. \tag{5.6}$$

For the plane stress or plane strain applications $f(\xi, \eta)$ would be $u(\xi, \eta)$, $v(\xi, \eta)$, $x(\xi, \eta)$ or $y(\xi, \eta)$.

5.3 Subroutine SFR2

We now describe a subroutine which calculates the shape functions $N_1(\xi, \eta), \ldots, N_8(\xi, \eta)$ and their derivatives $\partial N_1(\xi, \eta)/\partial \xi, \ldots, \partial N_8(\xi, \eta)/\partial \eta$ at any sampling point (ξ_P, η_P)—usually the Gauss point—within an element.

This routine can be considered as a standard routine and can be used for any formulation using a two-dimensional parabolic isoparametric element. In fact in the next chapter we shall use it in connection with the plate bending element.

Dictionary of variable names
DERIV(NDIME,NNODE) Shape function derivative at sampling point (ξ_P, η_P) within the element

$$\begin{bmatrix} \dfrac{\partial N_1}{\partial \xi}(\xi_P, \eta_P), \ldots, \dfrac{\partial N_8}{\partial \xi}(\xi_P, \eta_P) \\[3mm] \dfrac{\partial N_1}{\partial \eta}(\xi_P, \eta_P), \ldots, \dfrac{\partial N_8}{\partial \eta}(\xi_P, \eta_P) \end{bmatrix}$$

SHAPE(NNODE)   Shape functions associated with each node of current element sampled at (ξ_P, η_P)

$$\begin{bmatrix} N_1(\xi_P, \eta_P) \\ \vdots \\ N_8(\xi_P, \eta_P) \end{bmatrix}$$

S ξ_P—ξ coordinate of sampling point

T η_P—η coordinate of sampling point

Input/Output diagram

INPUT   OUTPUT

$A \begin{bmatrix} S \\ T \end{bmatrix} \rightarrow \boxed{\text{SFR2}} \rightarrow \begin{matrix} \text{SHAPE(NNODE)} \\ \text{DERIV(NDIME,NNODE)} \end{matrix} \Big]$ C(WORK)

Annotated FORTRAN *listing*

```
        SUBROUTINE SFR2(S,T)
C
C*** CALCULATES SHAPE FUNCTIONS AND THEIR
C    DERIVATIVES FOR 2D ELEMENTS
C
```

┌─────────────────────┐
│ COMMON BLOCKS   │
└─────────────────────┘

```
        S2=S*2.0
        T2=T*2.0
        SS=S*S
        TT=T*T
        ST=S*T
        SST=S*S*T
        STT=S*T*T
        ST2=S*T*2.0
```
⎫
⎬ 1 *
⎭

```
C
C*** SHAPE FUNCTIONS
C
        SHAPE(1)=(-1.0+ST+SS+TT-SST-STT)/4.0
        SHAPE(2)=(1.0-T-SS+SST)/2.0
        SHAPE(3)=(-1.0-ST+SS+TT-SST+STT)/4.0
        SHAPE(4)=(1.0+S-TT-STT)/2.0
        SHAPE(5)=(-1.0+ST+SS+TT+SST+STT)/4.0
        SHAPE(6)=(1.0+T-SS-SST)/2.0
        SHAPE(7)=(-1.0-ST+SS+TT+SST-STT)/4.0
        SHAPE(8)=(1.0-S-TT+STT)/2.0
```
⎫
⎬ 2 *
⎭

```
C
C*** SHAPE FUNCTION DERIVATIVES
C
        DERIV(1,1)=(T+S2-ST2-TT)/4.0
        DERIV(1,2)=-S+ST
        DERIV(1,3)=(-T+S2-ST2+TT)/4.0
        DERIV(1,4)=(1.0-TT)/2.0
```
⎫
⎬
⎭

```
DERIV(1,5)=(T+S2+ST2+TT)/4.0
DERIV(1,6)=-S-ST
DERIV(1,7)=(-T+S2+ST2-TT)/4.0
DERIV(1,8)=(-1.0+TT)/2.0
DERIV(2,1)=(S+T2-SS-ST2)/4.0
DERIV(2,2)=(-1.0+SS)/2.0
DERIV(2,3)=(-S+T2-SS+ST2)/4.0
DERIV(2,4)=-T-ST
DERIV(2,5)=(S+T2+SS+ST2)/4.0
DERIV(2,6)=(1.0-SS)/2.0
DERIV(2,7)=(-S+T2+SS-ST2)/4.0
DERIV(2,8)=-T+ST
RETURN
END
```
$\left.\vphantom{\begin{array}{c}1\\2\\3\\4\\5\\6\\7\\8\end{array}}\right\}3*$

1* Evaluate and store 2ξ, 2η, ξ^2, η^2, $\xi\eta$, $\xi^2\eta$, $\xi\eta^2$, $2\xi\eta$ at the Gauss point

2* Evaluate shape functions $N_1, N_2, \ldots, N_8$ at the Gauss point

3* Evaluate shape function derivatives $\partial N_1/\partial\xi$, $\partial N_1/\partial\eta, \ldots, \partial N_8/\partial\xi$, $\partial N_8/\partial\eta$ at the Gauss point

5.4 Jacobian matrix and the Cartesian shape function derivatives

The Jacobian matrix $\mathbf{J}(\xi, \eta)$ which will be required later is expressed, for two-dimensional situations, as

$$\mathbf{J} = \begin{bmatrix} \dfrac{\partial x}{\partial \xi} & \dfrac{\partial y}{\partial \xi} \\ \dfrac{\partial x}{\partial \eta} & \dfrac{\partial y}{\partial \eta} \end{bmatrix}$$

$$= \sum_{i=1}^{8} \begin{bmatrix} \dfrac{\partial N_i}{\partial \xi} \cdot x_i & \dfrac{\partial N_i}{\partial \xi} \cdot y_i \\ \dfrac{\partial N_i}{\partial \eta} \cdot x_i & \dfrac{\partial N_i}{\partial \eta} \cdot y_i \end{bmatrix}. \tag{5.7}$$

The inverse of the Jacobian matrix can be readily obtained using standard matrix inversion techniques

$$[\mathbf{J}]^{-1} = \begin{bmatrix} \dfrac{\partial \xi}{\partial x} & \dfrac{\partial \eta}{\partial x} \\ \dfrac{\partial \xi}{\partial y} & \dfrac{\partial \eta}{\partial y} \end{bmatrix} = \frac{1}{\det \mathbf{J}} \begin{bmatrix} \dfrac{\partial y}{\partial \eta} & -\dfrac{\partial y}{\partial \xi} \\ -\dfrac{\partial x}{\partial \eta} & \dfrac{\partial x}{\partial \xi} \end{bmatrix}. \tag{5.8}$$

To calculate the Cartesian derivatives of $f(\xi, \eta)$ we use the chain rule of differentiation

$$\frac{\partial f}{\partial x} = \frac{\partial f}{\partial \xi} \cdot \frac{\partial \xi}{\partial x} + \frac{\partial f}{\partial \eta} \cdot \frac{\partial \eta}{\partial x},$$

and   (5.9)

$$\frac{\partial f}{\partial y} = \frac{\partial f}{\partial \xi} \cdot \frac{\partial \xi}{\partial y} + \frac{\partial f}{\partial \eta} \cdot \frac{\partial \eta}{\partial y}.$$

For the plane stress element $f(\xi, \eta)$ can be $u(\xi, \eta)$ or $v(\xi, \eta)$.

It is important to note that an element of area $dx\, dy$ may be calculated by the expression

$$dx\, dy = \det \mathbf{J}\, d\xi\, d\eta,\qquad\qquad (5.10)$$

where $\det \mathbf{J}$ is the determinant of $\mathbf{J}$. Equation (5.10) will be referred to later when we consider the numerical integration of stiffness integrals.

5.5 Subroutine JACOB2

We now describe a subroutine which calculates:

the coordinates of the Gauss points

the Jacobian matrix

the inverse of the Jacobian matrix

the Cartesian shape function derivatives.

This routine can also be considered as a standard routine and can be used for any formulation using a two-dimensional parabolic isoparametric element. In the next chapter we shall use it again in connection with the plate bending element.

Dictionary of variable names

CARTD(NDIME,NNODE)   Cartesian shape function derivatives associated with the nodes of the current element sampled at any point (ξ_P, η_P) within the element

$$\begin{bmatrix} \dfrac{\partial N_1(\xi_P, \eta_P)}{\partial x}, \ldots, \dfrac{\partial N_8(\xi_P, \eta_P)}{\partial x} \\[2ex] \dfrac{\partial N_1(\xi_P, \eta_P)}{\partial y}, \ldots, \dfrac{\partial N_8(\xi_P, \eta_P)}{\partial y} \end{bmatrix}.$$

DJACB

Determinant of the Jacobian matrix sampled at any point (ξ_P, η_P) within the element

ELCOD(NDIME,NNODE)

Local array of nodal Cartesian coordinates of the element currently under consideration

$$\begin{bmatrix} x(\xi_1, \eta_1), \ldots, x(\xi_8, \eta_8) \\ y(\xi_1, \eta_1), \ldots, y(\xi_8, \eta_8) \end{bmatrix},$$

where ξ_i is the ξ coordinate of node i, etc.

GPCOD(NDIME,NGASP)

Local array of Cartesian coordinates of the Gauss points for element currently under consideration

$$\begin{bmatrix} x(\xi_{G_1}, \eta_{G_1}), \ldots, x(\xi_{G_8}, \eta_{G_8}), \ldots \\ y(\xi_{G_1}, \eta_{G_1}), \ldots, y(\xi_{G_8}, \eta_{G_8}), \ldots \end{bmatrix},$$

where $\xi_{G_i} = \xi$ coordinate of Gauss point G_i etc.

IELEM

Current element number

KGASP

Number of current Gauss point

SHAPE(NNODE)

Shape functions associated with the nodes of current element sampled at any point (ξ_P, η_P) within the element

$$\begin{bmatrix} N_1(\xi_P, \eta_P) \\ \vdots \\ N_8(\xi_P, \eta_P) \end{bmatrix}.$$

XJACM(NDIME,NDIME)

Jacobian matrix at sampling point

XJACI(NDIME,NDIME)

Inverse of Jacobian matrix at sampling point

DERIV (NDIME, NNODE), SHAPE (NNODE), IDIME, NDIME, INODE, NNODE are also used.

Input/Output diagram

INPUT	OUTPUT
C(WORK) $\begin{bmatrix} \text{DERIV(NDIME,NNODE)} \\ \text{SHAPE(NNODE)} \\ \text{ELCOD(NDIME,NNODE)} \end{bmatrix}$ → JACOB2 →	DJACB] A
A $\begin{bmatrix} \text{IELEM} \\ \text{KGASP} \end{bmatrix}$	CARTD(NDIME,NNODE)] C(WORK)
C(CONTRO) $\begin{bmatrix} \text{NDIME} \\ \text{NNODE} \end{bmatrix}$	

Annotated FORTRAN listing

```
      SUBROUTINE JACOB2(IELEM,DJACB,KGASP)
C
C*** CALCULATES COORDINATES OF GAUSS POINTS
C     AND THE JACOBIAN MATRIX AND ITS DETERMINANT
C     AND THE INVERSE FOR 2D ELEMENTS
C
      DIMENSION XJACM(2,2),XJACI(2,2)
```

┌─────────────────┐
│ COMMON BLOCKS │
└─────────────────┘

```
C
C*** CALCULATE COORDINATES OF SAMPLING POINT
C
      DO 10 IDIME=1,NDIME
      GPCOD(IDIME,KGASP)=0.0
      DO 10 INODE=1,NNODE                              ⎫
      GPCOD(IDIME,KGASP)=GPCOD(IDIME,KGASP)+           ⎬ 1*
    . ELCOD(IDIME,INODE)*SHAPE(INODE)                  ⎭
   10 CONTINUE
C
C*** CREATE JACOBIAN MATRIX XJACM
C
      DO 20 IDIME=1,NDIME                              ⎫
      DO 20 JDIME=1,NDIME                              ⎪
      XJACM(IDIME,JDIME)=0.0                           ⎬ 2*
      DO 20 INODE=1,NNODE                              ⎪
      XJACM(IDIME,JDIME)=XJACM(IDIME,JDIME)+           ⎪
    . DERIV(IDIME,INODE)*ELCOD(JDIME,INODE)            ⎭
   20 CONTINUE
C
C*** CALCULATE DETERMINANT AND INVERSE OF
C     JACOBIAN MATRIX
C
      DJACB=XJACM(1,1)*XJACM(2,2)-XJACM(1,2)*          ⎫ 3*
    . XJACM(2,1)                                       ⎭
      IF(DJACB.GT.0.0) GO TO 30                        ⎫
      WRITE(6,900) IELEM                               ⎬ 4*
      STOP                                             ⎭
   30 XJACI(1,1)=XJACM(2,2)/DJACB                      ⎫
      XJACI(2,2)=XJACM(1,1)/DJACB                      ⎪
      XJACI(1,2)=-XJACM(1,2)/DJACB                     ⎬ 5*
      XJACI(2,1)=-XJACM(2,1)/DJACB                     ⎭
C
C*** CALCULATE CARTESIAN DERIVATIVES
C
      DO 40 IDIME=1,NDIME                              ⎫
      DO 40 INODE=1,NNODE                              ⎪
      CARTD(IDIME,INODE)=0.0                           ⎪
      DO 40 JDIME=1,NDIME                              ⎬ 6*
      CARTD(IDIME,INODE)=CARTD(IDIME,INODE)+           ⎪
    . XJACI(IDIME,JDIME)*DERIV(JDIME,INODE)            ⎭
   40 CONTINUE
```

```
900 FORMAT(//,24HPROGRAM HALTED IN JACOB2,
   . /,11X,22H ZERO OR NEGATIVE AREA,/,
   . 10X,16H ELEMENT NUMBER ,I5)
    RETURN
    END
```

1* Calculate the coordinates of the Gauss point
2* Form the Jacobian matrix at the Gauss point as described in (5.7).
3* Calculate the determinant of the Jacobian matrix
4* If this determinant is less than or equal to zero print message and terminate execution of the program
5* Calculate the inverse of the Jacobian matrix
6* Calculate the Cartesian shape function derivatives as described in (5.9).

5.6 Strain matrix—B

In this section we discuss the formation of the strain matrix for the quadratic element.

The strain/displacement relationship for most elastic problems may be written in the form

$$\varepsilon = \mathbf{L}\delta, \tag{5.11}$$

where ε is the strain vector,

δ is the displacement vector,

$\mathbf{L}$ is the matrix of displacement differential operators.

Within an element we make the following approximation

$$\delta = \mathbf{N}\delta^e, \tag{5.12}$$

where $\mathbf{N}$ is the matrix of shape functions,

$$\mathbf{N} = [\mathbf{N}_1, \ldots, \mathbf{N}_8], \qquad \mathbf{N}_i = N_i\mathbf{I},$$

and $\mathbf{I}$ is an $n \times n$ identity matrix, where n is the number of degrees of freedom per node.

Using the finite element idealisation we can calculate strains from the following expression

$$\varepsilon = \mathbf{L}\mathbf{N}\delta^e = \mathbf{B}\delta^e$$
$$= [\mathbf{B}_1, \mathbf{B}_2, \ldots, \mathbf{B}_8]\delta^e, \tag{5.13}$$

where $\mathbf{B}$ is the element strain matrix.

In particular, for the plane stress/strain situations the strain/displacement relationship may be written as

$$
\begin{bmatrix} \dfrac{\partial u}{\partial x} \\[2mm] \dfrac{\partial v}{\partial y} \\[2mm] \dfrac{\partial u}{\partial y} + \dfrac{\partial v}{\partial x} \end{bmatrix} = \begin{bmatrix} \dfrac{\partial}{\partial x} & 0 \\[2mm] 0 & \dfrac{\partial}{\partial y} \\[2mm] \dfrac{\partial}{\partial y} & \dfrac{\partial}{\partial x} \end{bmatrix} \begin{bmatrix} u \\ v \end{bmatrix} \tag{5.14}
$$

$$ \varepsilon = \mathbf{L}\delta. $$

Using the finite element idealisation we can write

$$
\varepsilon = \sum_{i=1}^{8} \begin{bmatrix} \dfrac{\partial N_i}{\partial x} & 0 \\[2mm] 0 & \dfrac{\partial N_i}{\partial y} \\[2mm] \dfrac{\partial N_i}{\partial y} & \dfrac{\partial N_i}{\partial x} \end{bmatrix} \begin{bmatrix} u_i \\ v_i \end{bmatrix} = \sum_{i=1}^{8} \mathbf{B}_i \delta_i. \tag{5.15}
$$

Since the calculation of the Cartesian shape function derivatives has already been dealt with, the calculation of the strain matrix **B** can be simply achieved by assembling these quantities into their correct position in the strain matrix.

5.7 Subroutine BMATPS

This subroutine calculates the strain matrix **B** for plane stress and plane strain problems using the Cartesian shape function derivatives.

Dictionary of variable names
BMATX(NSTRE,NEVAB) The element strain matrix at any point within the element

$$ \mathbf{B} = [\mathbf{B}_1, \mathbf{B}_2, \ldots, \mathbf{B}_8]. $$

CARTD(NDIME,NNODE),NNODE are also used.

Input/Output diagram
             INPUT   OUTPUT

C(WORK) [CARTD(NDIME,NNODE) → | BMATPS | → BMATX(NSTRE,NEVAB)] C(WORK)

Annotated FORTRAN *listing*

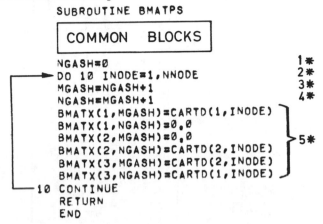

```
            SUBROUTINE BMATPS

      ┌─────────────────────────┐
      │  COMMON    BLOCKS        │
      └─────────────────────────┘
         NGASH=0                                          1*
    ┌──► DO 10 INODE=1,NNODE                              2*
    │    MGASH=NGASH+1                                    3*
    │    NGASH=MGASH+1                                    4*
    │    BMATX(1,MGASH)=CARTD(1,INODE) ┐
    │    BMATX(1,NGASH)=0.0            │
    │    BMATX(2,MGASH)=0.0            │
    │    BMATX(2,NGASH)=CARTD(2,INODE) ├5*
    │    BMATX(3,MGASH)=CARTD(2,INODE) │
    │    BMATX(3,NGASH)=CARTD(1,INODE) ┘
    └─ 10 CONTINUE
         RETURN
         END
```

1* Initialise NGASH which is a counter used to locate positions in the **B** matrix

2* Loop for each node of the element

3* Find column position in **B** associated with u_i

4* Find column position in **B** associated with v_i

5* Complete $\mathbf{B}_i$ matrix where $\mathbf{B} = [\mathbf{B}_1, \ldots, \mathbf{B}_8]$ and

$$\mathbf{B}_i = \begin{bmatrix} \dfrac{\partial N_i}{\partial x} & 0 \\[3mm] 0 & \dfrac{\partial N_i}{\partial y} \\[3mm] \dfrac{\partial N_i}{\partial y} & \dfrac{\partial N_i}{\partial x} \end{bmatrix}.$$

5.8 The matrix of elastic constants—D

The stress/strain relationship for an elastic material, in the absence of initial stresses and strains, may be written in the form

$$\sigma = \mathbf{D}\varepsilon, \tag{5.16}$$

where **D** is the matrix of elastic constants. For plane stress situations and assuming isotropic materials

$$\mathbf{D} = \frac{E}{1 - v^2} \begin{bmatrix} 1 & v & 0 \\ v & 1 & 0 \\ 0 & 0 & \frac{1 - v}{2} \end{bmatrix}, \tag{5.17}$$

whereas for plane strain situations (isotropic case)

$$\mathbf{D} = \frac{E(1 - v)}{(1 + v)(1 - 2v)} \begin{bmatrix} 1 & \frac{v}{1 - v} & 0 \\ \frac{v}{1 - v} & 1 & 0 \\ 0 & 0 & \frac{1 - 2v}{2(1 - v)} \end{bmatrix}. \tag{5.18}$$

These expressions can be verified from any standard elasticity text [2].

5.9 Subroutine MODPS

This routine simply evaluates the elasticity matrix **D** for either plane stress or plane strain applications.

Dictionary of variable names

DMATX(NSTRE,NSTRE)	The elasticity matrix **D**
LPROP	Element property type
NTYPE	The problem type parameter $\begin{cases} 1 = \text{plane stress} \\ 2 = \text{plane strain} \end{cases}$
POISS	Poisson's ratio (local name)
PROPS(LPROP,1)	Young's modulus
PROPS(LPROP,2)	Poisson's ratio
YOUNG	Young's modulus (local name)

Input/Output diagram

A [LPROP   → | MODPS | → DMATX(NSTRE,NSTRE)] C(WORK)

C(LGDATA) [PROPS(NMATS,NPROP)

C(CONTRO) [NSTRE

Annotated FORTRAN listing

```
            SUBROUTINE MODPS(LPROP)

       ┌─────────────────────────┐
       │   COMMON BLOCKS         │
       └─────────────────────────┘
            YOUNG=PROPS(LPROP,1)                                    } 1*
            POISS=PROPS(LPROP,2)
      ┌───► DO 10 ISTRE=1,NSTRE
    ┌─┼───► DO 10 JSTRE=1,NSTRE                                     } 2*
    │ │     DMATX(ISTRE,JSTRE)=0.0
    └─┴─ 10 CONTINUE
            IF(NTYPE.NE.1) GO TO 20                                   3*
      C
      C*** D MATRIX FOR PLANE STRESS CASE
      C
            CONST=YOUNG/(1.0-POISS*POISS)                             4*
            DMATX(1,1)=CONST
            DMATX(2,2)=CONST
            DMATX(1,2)=CONST*POISS                                  } 5*
            DMATX(2,1)=CONST*POISS
            DMATX(3,3)=(1.0-POISS)*CONST/2.0
            GO TO 30
         20 IF(NTYPE.NE.2) GO TO 30
      C
      C*** D MATRIX FOR PLANE STRAIN CASE
      C
            CONST=YOUNG*(1.0-POISS)/((1.0+POISS)*
          . (1.0-2.0*POISS))
            DMATX(1,1)=CONST
            DMATX(2,2)=CONST
            DMATX(1,2)=CONST*POISS/(1.0-POISS)                     } 6*
            DMATX(2,1)=CONST*POISS/(1.0-POISS)
            DMATX(3,3)=CONST*(1.0-2.0*POISS)/
          . (2.0*(1.0-POISS))
         30 CONTINUE
            RETURN
            END
```

1* Determine the elastic modulus and Poisson's ratio for the element currently under consideration
2* Zero the **D** matrix
3* If parameter NTYPE is not equal to 1 jump to statement 20. In other words, are we considering a plane stress or a plane strain situation?
4* Evaluate $E/(1 - v^2)$
5* Evaluate the **D** matrix for the plane stress situation
6* Evaluate the **D** matrix for the plane strain situation

5.10 Stress matrix S

Since the strain/displacement relationship in the finite element approximation may be written as

$$\varepsilon = [\mathbf{B}_1, \ldots, \mathbf{B}_8]\delta^e \tag{5.19}$$

the stress at any point within the element can be expressed as

$$\sigma = \mathbf{DB}\delta^e = \mathbf{S}\delta^e \tag{5.20}$$

Thus for a plane stress situation we have

$$\sigma = \begin{bmatrix} \sigma_x \\ \sigma_y \\ \tau_{xy} \end{bmatrix} = \sum_{i=1}^{8} \frac{E}{1-v^2} \begin{bmatrix} 1 & v & 0 \\ v & 1 & 0 \\ 0 & 0 & \frac{1-v}{2} \end{bmatrix} \begin{bmatrix} \frac{\partial N_i}{\partial x} & 0 \\ 0 & \frac{\partial N_i}{\partial y} \\ \frac{\partial N_i}{\partial y} & \frac{\partial N_i}{\partial x} \end{bmatrix} \begin{bmatrix} u_i \\ v_i \end{bmatrix} \tag{5.21}$$

A similar expression can be obtained for the plane strain condition if we substitute the appropriate elasticity matrix from (5.18) into (5.21).

5.11 Subroutine DBE

The routine which multiplies matrix $\mathbf{D}$ by matrix $\mathbf{B}$ is identical to that used for the beam element and since it is described in Section 4.10.6 it will not be repeated here. N.B. The product $\mathbf{DB}$ is stored in DBMAT(ISTRE,IEVAB) where ISTRE ranges over the number of in-plane stress components ($=3$ for plane applications) and IEVAB ranges over the element variables.

5.12 The element stiffness matrix

We now have all of the information necessary to calculate the element stiffness matrix $\mathbf{K}^e$ where, from (1.28),

$$\mathbf{K}^e = \iint [\mathbf{B}]^T \mathbf{DB} \, dv. \tag{5.22}$$

In fact, a typical submatrix of $\mathbf{K}^e$ linking nodes i and j may be evaluated from the expression

$$\mathbf{K}_{ij}^e = \iint [\mathbf{B}_i]^T \mathbf{DB}_j t \, \det \mathbf{J} \, d\xi \, d\eta, \tag{5.23}$$

where t is the element thickness and

$$dx \, dy = \det \mathbf{J} \, d\xi \, d\eta.$$

As with the beam element we use numerical integration to evaluate these stiffness integrals. Again Gaussian quadrature is adopted, but because we are integrating over an area instead of along a line we must use product rules. The two-dimensional case can be obtained by combining one-dimensional formulae.

For example consider the following integral which is to be evaluated using a 3-point rule in both the ξ and η directions.

$$
\begin{aligned}
I_{3,3} &= \int_{-1}^{+1} \int_{-1}^{+1} \phi(\xi, \eta)\, d\xi\, d\eta = \int_{-1}^{+1} \left\{ \int_{-1}^{+1} \phi(\xi, \eta)\, d\xi \right\} d\eta \\
&= \int_{-1}^{+1} (a_I \phi(\xi_I, \eta) + a_{II} \phi(\xi_{II}, \eta) + a_{III} \phi(\xi_{III}, \eta))\, d\eta \\
&= a_I \{ a_I \phi(\xi_I, \eta_I) + a_{II} \phi(\xi_I, \eta_{II}) + a_{III} \phi(\xi_1, \eta_{III}) \} \\
&\quad + a_{II} \{ a_I \phi(\xi_{II}, \eta_I) + a_{II} \phi(\xi_{II}, \eta_{II}) + a_{III} \phi(\xi_{II}, \eta_{III}) \} \\
&\quad + a_{III} \{ a_I \phi(\xi_{III}, \eta_I) + a_{II} \phi(\xi_{III}, \eta_{II}) + a_{III} \phi(\xi_{III}, \eta_{III}) \},
\end{aligned}
\tag{5.24}
$$

where ξ_i is the ξ coordinate of the ith Gauss point

η_i is the η coordinate of the ith Gauss point

a_i is the weighting factor.

Subroutine GAUSSQ (described in Chapter 3) sets up the sampling point positions and weighting factors for numerical integration where according to the information provided only 2-point and 3-point rules are allowed. allowed.

5.13 Subroutine STIFPS

This routine which calculates the element stiffness and stress matrices also contains all the features of routine STIFB which calculates the stiffness matrix, etc., for the beam element. The element stiffness matrix is defined by (1.28) where it is seen that this matrix depends entirely on the previously assembled **B** and **D** matrices. For the present application the stiffness matrix is square and symmetric and of size NEVAB × NEVAB where NEVAB = NNODE × NDOFN and it is recalled that NNODE defines the number of nodes per element (8 for the parabolic isoparametric element) and NDOFN is the number of degrees of freedom per node (2 for this case). This element stiffness matrix will be temporarily stored in the array

ESTIF(NEVAB,NEVAB)

before transfer to disc file 1 for later use in the solution routine. The matrix product **DB** for each element is calculated for future use in the evaluation of stresses according to (1.32). This matrix product for each Gauss point will first be temporarily stored in the array

$$\text{DBMAT(NSTRE,NEVAB)}$$

The values are then transferred to the array

$$\text{SMATX(NSTRE,NEVAB,KGASP)}$$

which contains the stress matrix for each Gauss point in an element. As soon as the matrix product has been evaluated for each Gauss point of the element the values are transferred to disc file 3. The Gauss point coordinates for each Gauss point within an element are also calculated and stored in the array

$$\text{GPCOD(NDIME,KGASP)}$$

This information is then also transferred to disc file 3.

Input/Output diagram

Annotated FORTRAN *listing*

```
          SUBROUTINE STIFPS
          DIMENSION ESTIF(16,16)
```

```
          COMMON BLOCKS
```

```
     C
     C*** LOOP OVER EACH ELEMENT
     C
             DO 70 IELEM=1,NELEM                          1*
             LPROP=MATNO(IELEM)                           2*
     C
     C*** EVALUATE THE COORDINATES OF THE ELEMENT
```

```
C     NODAL POINTS
C
   ┌──── DO 10 INODE=1,NNODE
   │     LNODE=LNODS(IELEM,INODE)          ⎫
   │ ┌── DO 10 IDIME=1,NDIME                ⎬ 3*
   └─10 ELCOD(IDIME,INODE)=COORD(LNODE,IDIME) ⎭
C
C*** EVALUATE THE D-MATRIX
C
      CALL MODPS(LPROP)                      4*
      THICK=PROPS(LPROP,3)                   5*
C
C*** INITIALIZE THE ELEMENT STIFFNESS MATRIX
C
   ┌──── DO 20 IEVAB=1,NEVAB                 ⎫
   │ ┌── DO 20 JEVAB=1,NEVAB                 ⎬ 6*
   └─20 ESTIF(IEVAB,JEVAB)=0.0               ⎭
      KGASP=0                                7*
C
C*** ENTER LOOPS FOR AREA NUMERICAL INTEGRATION
C
   ┌──── DO 50 IGAUS=1,NGAUS                 ⎫ 8*
   │ ┌── DO 50 JGAUS=1,NGAUS                 ⎭
   │ │   KGASP=KGASP+1                       9*
   │ │   EXISP=POSGP(IGAUS)                  ⎫
   │ │   ETASP=POSGP(JGAUS)                  ⎬10*
C
C*** EVALUATE THE SHAPE FUNCTIONS,ELEMENTAL
C     VOLUME,ETC,
C
      CALL SFR2(EXISP,ETASP)                 11*
      CALL JACOB2(IELEM,DJACB,KGASP)         12*
      DVOLU=DJACB*WEIGP(IGAUS)*WEIGP(JGAUS)  13*
      IF(THICK.NE.0.0) DVOLU=DVOLU*THICK     14*
C
C*** EVALUATE THE B AND DB MATRICES
C
      CALL BMATPS                            15*
      CALL DBE                               16*
C
C*** CALCULATE THE ELEMENT STIFFNESSES
C
   ┌──── DO 30 IEVAB=1,NEVAB                 ⎫
   │ ┌── DO 30 JEVAB=IEVAB,NEVAB            ⎬17*
   │ │ ┌ DO 30 ISTRE=1,NSTRE                ⎭
   └─30 ESTIF(IEVAB,JEVAB)=ESTIF(IEVAB,JEVAB)+ ⎫
      , BMATX(ISTRE,IEVAB)*DBMAT(ISTRE,      ⎬18*
      , JEVAB)*DVOLU                         ⎭
C
C*** STORE THE COMPONENTS OF THE DB MATRIX FOR
C     THE ELEMENT
C
   ┌──── DO 40 ISTRE=1,NSTRE                 ⎫
   │ ┌── DO 40 IEVAB=1,NEVAB                 ⎬19*
   └─40 SMATX(ISTRE,IEVAB,KGASP)=DBMAT(ISTRE,IEVAB) ⎭
   └──50 CONTINUE                            20*
C
```

```
C*** CONSTRUCT THE LOWER TRIANGLE OF THE
C    STIFFNESS MATRIX
C
        DO 60 IEVAB=1,NEVAB
        DO 60 JEVAB=1,NEVAB                              } 21*
   60 ESTIF(JEVAB,IEVAB)=ESTIF(IEVAB,JEVAB)
C
C*** STORE THE STIFFNESS MATRIX,STRESS MATRIX
C    AND SAMPLING POINT COORDINATES FOR EACH
C    ELEMENT ON DISC FILE
C
        WRITE(1) ESTIF                                    22*
        WRITE(3) SMATX,GPCOD                              23*
   70 CONTINUE                                            24*
      RETURN
      END
```

1* Loop over each element
2* Identify the material property of the element
3* Store the coordinates of the nodal points of the element under consideration in the local array ELCOD(NDIME,NNODE)
4* Call subroutine MODPS(LPROP) which sets up the **D** matrix for the material. Different entries are made for plane stress and plane strain as governed by the index NTYPE input in subroutine INPUT
5* Set THICK equal to the element thickness
6* Set the element stiffness array ESTIF(NEVAB,NEVAB) to zero
7* Set index KGASP to zero. This counter will be used to indicate the Gauss point number of an element
8* Enter loops for the numerical integration required in (1.28)
9* Increment KGASP by one to give the current Gauss point number
10* Set up the coordinates ξ_n, η_m of the Gaussian sampling points for use in (1.28)
11* Evaluate the shape functions N_i and the derivatives $\partial N_i/\partial \xi$, $\partial N_i/\partial \eta$ corresponding to ξ_n, η_m
12* Calculate the Jacobian Matrix **J**, its determinant, DJACB, its inverse $[\mathbf{J}]^{-1}$ and the Cartesian derivatives of the shape functions. This subroutine also calculates the Cartesian coordinates x, y of the Gauss points which are stored in array GPCOD
13* Calculate the elemental volume DVOLU for numerical integration
14* If the material thickness is specified as a non-zero quantity multiply DVOLU by the thickness
15* Call subroutine BMATPS which generates the **B** matrix for the current Gauss point
16* Call subroutine DBE which forms the matrix product **DB** for the current Gauss point
17* Enter two loops over the range NEVAB where NEVAB is the total

number of degrees of freedom per element. These loops control the
construction of the element stiffness matrix ESTIF(NEVAB,NEVAB)

18* Construct the element stiffness matrix according to (1.28). It should
be noted that only the upper triangle is formed at this stage due to
DO LOOP parameters in 17*

19* Store the matrix product **DB** for each Gaussian sampling point in the
array SMATX(NSTRE,NEVAB,KGASP)

20* Termination of loop for numerical integration

21* Complete the element stiffness matrix by symmetry

22* Write onto a disc file, the element stiffness matrix ESTIF(NEVAB,
NEVAB)

23* Also write the stress matrix SMATX(NSTRE,NEVAB,KGASP) and
the Gauss point coordinates GPCOD(NDIME,KGASP) to another
disc file

24* Termination of DO LOOP over each element

The element stiffnesses are then entered into the solution routine described
in Chapter 8 where the nodal displacements are determined, output and
stored. It then remains to calculate the element stresses.

5.14 Calculation of element stresses

As explained in Chapter 3 it is essential that element stresses be evaluated
at points within elements in view of the possible large area covered by a
single element. Since the Gaussian sampling points within each element have
already been defined, it is convenient to output the components of stress at
these points.

The stresses are readily calculated from the displacements by use of (1.32).
For this purpose the matrix product **DB** has already been evaluated in Section
5.11 at each Gaussian point of each element and stored on disc file 3.

On exit from solution routine FRONT the displacements of each nodal
point will have been stored in the array

$$ASDIS(ITOTV)$$

where ITOTV ranges from 1 to the total number of degrees of freedom
for the whole structure.

As each element is processed separately, it is convenient to store displacements associated with the nodal points of each element in turn temporarily
in the local array

$$ELDIS(IDOFN,INODE)$$

where IDOFN ranges from 1 to the total number of degrees of freedom per node and INODE ranges from 1 to the total number of nodes per element.

The stress components at each element Gauss point will be calculated and stored in the local array

$$\text{STRSG(ISTRE)}$$

where ISTRE ranges from 1 to 4, denoting the four relevant stress components, (since for plane strain problems, the through thickness stress, σ_z is non zero). For plane strain problems the through-thickness stress is given by the expression

$$\sigma_z = v(\sigma_x + \sigma_y), \tag{5.25}$$

where v is the Poisson's ratio and σ_x, σ_y are the in-plane normal stress components.

For plane stress problems the through thickness stress σ_z is zero. If thermal effects are present then, according to (1.32) initial stresses of $-\mathbf{D}\varepsilon^0$ must be added to those calculated from the displacements. These will be evaluated at each element Gauss point in subroutine LOADPS formulated in Chapter 7 and stored in the array

$$\text{STRIN(ISTR1,KGAST)}$$

where ISTR1 ranges over the four stress components and the subscript KGAST defines the element Gauss point number. In addition to the Cartesian stress components the principal stress values will also be calculated. The expression for the principal stresses can be found in any standard text on the theory of elasticity (e.g. [2, 3]) and are given below

$$\sigma_1 = \frac{\sigma_x + \sigma_y}{2} + \sqrt{\frac{(\sigma_x - \sigma_y)^2}{4} + \tau_{xy}^2}$$

$$\sigma_2 = \frac{\sigma_x + \sigma_y}{2} - \sqrt{\frac{(\sigma_x - \sigma_y)^2}{4} + \tau_{xy}^2}$$

$$\alpha = \tfrac{1}{2}\text{Tan}^{-1}\left(\frac{2\tau_{xy}}{\sigma_x - \sigma_y}\right), \tag{5.26}$$

where σ_1 and σ_2 are respectively the maximum and minimum principal stress and α is the angle which the maximum principal stress makes with the positive x axis. The principal stress values will be stored in the array

$$\text{STRSP(ISTRE)}$$

where ISTRE ranges from 1 to 3. The first two values will contain the maximum and minimum principal stresses respectively and the third value will store the angle α. It should be mentioned that only the principal value of the angle (i.e. $-90° \leqslant \alpha \leqslant 90°$) will be output. The subroutine for the evaluation of stresses can now be assembled.

5.15 Subroutine STREPS

This subroutine is called after the displacements have been evaluated and calculates the stresses at the Gauss points.

Input/Output diagram

```
              INPUT                              OUTPUT

          ┌ NELEM                    →│ STREPS │→ STRSG(NSTR1)        ┐
          │ NNODE                                STRSP(NSTRE)         │ F(6)
          │ NDOFN                                GPCOD(NDIME,NGASP)   ┘
C(CONTRO) │ NGAUS
          │ NSTRE
          └ NDIME

          ┌ ASDIS(NPOIN × NDOFN)
C(LGDATA) └ STRIN(NSTRE,NELEM × NGASP)

          ┌ SMATX(NSTRE,NEVAB,NGASP)
F(3)      └ GPCOD(NDIME,NGASP)
```

Annotated FORTRAN *listing*

```
      SUBROUTINE STREPS
      DIMENSION STRSP(3),ELDIS(2,8),STRSG(4)
```

```
┌─────────────────────────┐
│    COMMON  BLOCKS        │
└─────────────────────────┘
```

```
      NSTR1=NSTRE+1                                          1*
      WRITE(6,900)                                        ┐
      WRITE(6,905)                                        ┘  2*
  905 FORMAT(1H0,4HG,P,,2X,8HX=COORD,,2X,
     . 8HY=COORD,,3X,8HX-STRESS,4X,8HY=STRESS,
     . 3X,9HXY=STRESS,3X,8HZ=STRESS,4X,
     . 8HMAX P.S.,4X,8HMIN &.S.,6X,5HANGLE)
      KGAST=0                                                3*
    C
    C*** LOOP OVER EACH ELEMENT
    C
┌─────── DO 60 IELEM=1,NELEM                                 4*
│         LPROP=MATNO(IELEM)                                 5*
│         POISS=PROPS(LPROP,2)                               6*
│       C
│       C*** READ THE STRESS MATRIX , SAMPLING POINT
│       C    COORDINATES FOR THE ELEMENT
│       C
│         READ(3) SMATX,GPCOD                                7*
│         WRITE(6,910) IELEM                                 8*
```

```
C
C*** IDENTIFY THE DISPLACEMENTS OF THE
C    ELEMENT NODAL POINTS
C
      DO 10 INODE=1,NNODE
      LNODE=LNODS(IELEM,INODE)
      NPOSN=(LNODE-1)*NDOFN
      DO 10 IDOFN=1,NDOFN                          } 9*
      NPOSN=NPOSN+1
      ELDIS(IDOFN,INODE)=ASDIS(NPOSN)
   10 CONTINUE
      KGASP=0                                       10*
C
C*** ENTER LOOPS OVER EACH SAMPLING POINT
C
      DO 50 IGAUS=1,NGAUS                          } 11*
      DO 50 JGAUS=1,NGAUS
      KGAST=KGAST+1                                  12*
      KGASP=KGASP+1                                  13*
C
C*** COMPUTE THE CARTESIAN STRESS COMPONENTS
C    AT THE SAMPLING POINTS
C
      DO 20 ISTRE=1,NSTRE
      STRSG(ISTRE)=0.0
      KGASH=0
      DO 20 INODE=1,NNODE
      DO 20 IDOFN=1,NDOFN                          } 14*
      KGASH=KGASH+1
      STRSG(ISTRE)=STRSG(ISTRE)+SMATX(ISTRE,
    . KGASH,KGASP)*ELDIS(IDOFN,INODE)
   20 CONTINUE
C
C*** COMPUTE THE OUT OF PLANE NORMAL STRESS
C    COMPONENT
C
      IF(NTYPE.EQ.2) STRSG(4)=POISS*(STRSG(1)+
    . STRSG(2))                                    } 15*
      IF(NTYPE.EQ.1) STRSG(4)=0.0
C
C*** FOR THERMAL LOADING ADD ON THE INITIAL
C    THERMAL STRESSES
C
      IF(ITEMP.EQ.0) GO TO 40                        16*
      DO 30 ISTR1=1,NSTR1
      STRSG(ISTR1)=STRSG(ISTR1)+STRIN(ISTR1,KGAST) } 17*
   30 CONTINUE
C
C*** COMPUTE THE PRINCIPAL STRESSES
C
   40 XGASH=(STRSG(1)+STRSG(2))*0.5
      XGISH=(STRSG(1)-STRSG(2))*0.5
      XGESH=STRSG(3)
      XGOSH=SQRT(XGISH*XGISH+XGESH*XGESH)          } 18*
```

```
        STRSP(1)=XGASH+XGOSH
        STRSP(2)=XGASH-XGOSH
        IF(XGISH.EQ.0.0) XGISH=0.1E-20
        STRSP(3)=ATAN(XGESH/XGISH)*28.647889757
C
C*** OUTPUT THE STRESSES
C
        WRITE(6,915) KGASP,(GPCOD(IDIME,KGASP),
       .IDIME=1,NDIME),(STRSG(ISTR1),ISTR1=1,NSTR1),        19*
       .(STRSP(ISTRE),ISTRE=1,NSTRE)
   50 CONTINUE                                              20*
   60 CONTINUE                                              21*
  900 FORMAT(/,10X,8HSTRESSES,/)
  910 FORMAT(/,5X,12HELEMENT NO.=,I5)
  915 FORMAT(I5,2F10.4,6E12.5,F10.4)
      RETURN
      END
```

1* Preset the number of stresses to 4 to allow for through-thickness component

2* Write heading and titles for stresses

3* Set KGAST=0. This index is used to identify the Gauss point number for thermal loading where a running total is kept from 1 to the total number of Gauss points in the structure

4* Loop over each element

5* Extract material property set number of current element

6* Calculate Poisson's ratio for current element

7* Read from file 3 the matrix product **DB** and Gauss point coordinate for each Gauss point of the current element

8* Write heading naming current element

9* Store the displacements of the nodal points associated with the current element in the local array ELDIS(NDIME,NNODE)

10* Set KGASP=0. This index is used to identify the Gauss point number for use in the SMATX array. Here, as soon as an element has been completely processed KGASP is reset to zero. This differs from the thermal case where a running total over all elements is kept

11* Enter loops for processing each element Gauss point. It should be noted that, according to (1.32), no numerical integration is necessary

12* Increment KGAST by one to give the location of the current Gauss point for use in the STRIN array

13* Increment KGASP by one to give the location of the current Gauss point for use in the SMATX array

14* Evaluate the first term in (1.32)

i.e. $$\boldsymbol{\sigma} = \mathbf{DB}\boldsymbol{\delta}^e$$

15* Calculate the stress component in the through-thickness direction for either plane stress or plane strain problems as governed by the index NTYPE

16* If no thermal effects are to be considered go to 40

17* For thermal problems add the term $-\mathbf{D}\varepsilon^0$ to the values calculated in 14* according to (1.32)

18* Evaluate the principal stresses and their direction according to (5.26)

19* Output the Cartesian coordinates of the Gauss points, the Cartesian stress components, the principal stresses and their direction. The order in which the Gauss point stresses are output must now be given detailed attention. In all loops over the element Gauss points in the program the order of looping is as indicated in 11*, i.e. the inner loop is always the one associated with the η local coordinate. Consequently for a three point Gauss rule, for example, the order of numbering for an element is as shown in Fig. 5.2, being controlled by the local coordinate directions ξ, η which in turn depend upon the sequence in which the element nodal point numbers are specified as described in Section 5.2.

20* End of Gauss loop

21* End of element loop

References

1. Turner, M. J., Clough, R. W., Martin, H. C., and Topp, L. J., Stiffness and deflection analysis of complex structures. *J. Aero. Sci.* **23**, 805–823, 1956.
2. Timoshenko, S., and Goodier, J. N., "Theory of Elasticity". McGraw-Hill, New York, 1951.
3. Wang, C. T., "Applied Elasticity". McGraw-Hill, New York, 1953.

6

Element Characteristics for Plate Bending Applications

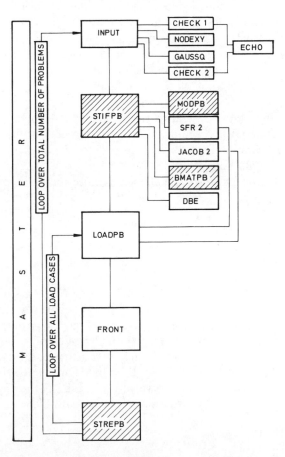

6.1 Introduction and basic theory

The classical problems of plate flexure were amongst the first subjects for the application of the finite element method. Primary attention has been given to solving problems based on the so-called 'thin plate' theory in which shear deformations are neglected. Even here considerable difficulties were encountered in the displacement type formulation due to the requirement of slope continuity between adjacent elements and many early attempts violated this condition while others introduced very complex functions for its satisfaction. A full history of the developments is given in [1], for example, and need not be repeated here except to say that two of the early non-conforming elements [2], [3] (i.e. violating the continuity requirements) have been proved convergent under most practical conditions and are today widely used for the design and analysis of slabs despite the later developments of more sophisticated approaches.

When shear deformation is of importance—as it is in thick plates, cellular plates and plates of sandwich construction—the solution to the problem becomes significantly more complex and much effort has recently been expended in finding an adequate approach. In this chapter, an approach is presented which not only permits *all types of plates to be accommodated in the solution scheme* but also avoids the continuity requirements which have made the solution of thin plates so difficult. Again, for consistency, an 8-noded isoparametric element is used.

The reader may find it useful to review Chapter 4 on the isoparametric beam element before proceeding with the theoretical aspects of the plate element. The beam element may be considered to be a one-dimensional equivalent of the plate element.

In the formulation of the plate element we use the assumptions adopted by Mindlin [4]:

the deflections of the plate (w) are small

normals to the midsurface before deformation remain straight but not necessarily normal to the midsurface after deformation

stresses normal to the midsurface are negligible irrespective of the loading.

The displacement field can thus be uniquely specified by an independent variation of the lateral displacement w and of the two angles defining the direction of the line originally normal to the midsurface of the plate as shown in Fig. 6.1. The variables θ_x and θ_y can therefore be considered as average rotations and a correction will be made subsequently to allow for non-uniform shear distribution (or warping). In Fig. 6.1 the angles ϕ_x and ϕ_y denote the average shear deformations and for both the x and y directions

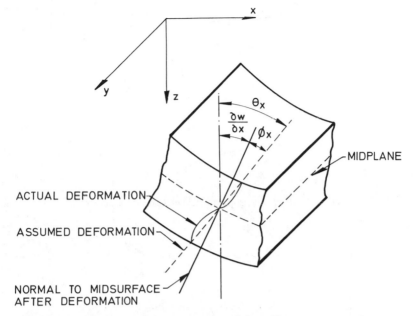

FIG. 6.1. Deformation of the cross-section of plate of homogeneous section.

we can write

$$\boldsymbol{\delta} = \begin{bmatrix} w \\ \theta_x \\ \theta_y \end{bmatrix} = \begin{bmatrix} w \\ \dfrac{\partial w}{\partial x} + \phi_x \\ \dfrac{\partial w}{\partial y} + \phi_y \end{bmatrix},$$

(6.1)

and

$$\boldsymbol{\phi} = \begin{bmatrix} -\phi_x \\ -\phi_y \end{bmatrix}.$$

The x–y plane is taken to coincide with the midsurface of the plate as shown in Fig. 6.2 and the plate thickness is denoted by t. The external forces are considered to be applied at the midsurface and we work with two-dimensional force quantities, i.e. body forces per unit area. Here we are using the term "force" in the generalised sense to denote either an actual force (in the z direction) or moments (in the xz or yz planes).

The bending moments and shear forces are shown in Fig. 6.3 and they are

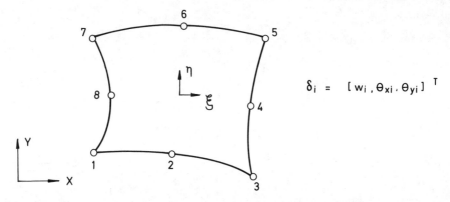

$$\delta_i = [w_i, \theta_{xi}, \theta_{yi}]^T$$

FIG. 6.2.   Parabolic isoparametric plate bending element.

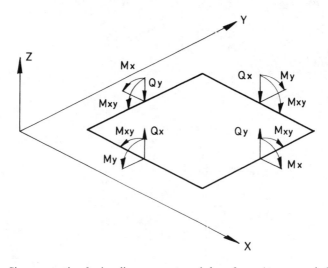

FIG. 6.3.   Sign convention for bending moments and shear forces ($+$ ve sense indicated).

defined by the expressions

$$\mathbf{M} = \begin{bmatrix} M_x \\ M_y \\ M_{xy} \end{bmatrix} = \begin{bmatrix} \int \sigma_x z \, dz \\ \int \sigma_y z \, dz \\ \int \tau_{xy} z \, dz \end{bmatrix},$$

and   (6.2)

$$\mathbf{Q} = \begin{bmatrix} Q_x \\ Q_y \end{bmatrix} = \begin{bmatrix} \int \tau_{xz} \, dz \\ \int \tau_{yz} \, dz \end{bmatrix},$$

where the above integrals are calculated over the range $z = -t/2$ to $z = +t/2$. Thus the total potential energy of the plate may be written as

$$\pi = \frac{1}{2} \int_A (M_x \chi_x + M_y \chi_y + M_{xy} \chi_{xy} + Q_x \phi_x + Q_y \phi_y)\, dA - \int_A qw\, dA,$$

(6.3)

where χ_x, χ_y, χ_{xy} are measures of bending deformation

$$\chi = \begin{bmatrix} \chi_x \\ \chi_y \\ \chi_{xy} \end{bmatrix} = \begin{bmatrix} -\dfrac{\partial \theta_x}{\partial x} \\ -\dfrac{\partial \theta_y}{\partial y} \\ -\left(\dfrac{\partial \theta_x}{\partial y} + \dfrac{\partial \theta_y}{\partial x}\right) \end{bmatrix},$$

(6.4)

and q is the distributed load per unit area.

Thus the total potential energy π may be written in an abbreviated form

$$\pi = \frac{1}{2} \int_A ([M]^T \chi + [Q]^T \phi)\, dA - \int_A qw\, dA.$$

(6.5)

The stress–strain relationships may be written in the following form

$$M = D_f \chi,$$

and

(6.6)

$$Q = D_s \phi,$$

where for an isotropic homogeneous material

$$D_f = \frac{Et^3}{12(1 - v^2)} \begin{bmatrix} 1 & v & 0 \\ v & 1 & 0 \\ 0 & 0 & (1 - v)/2 \end{bmatrix},$$

(6.7)

and

$$D_s = \frac{Et}{2(1 + v)\alpha} \begin{bmatrix} 1 & 0 \\ 0 & 1 \end{bmatrix},$$

where $\alpha = \frac{6}{5}$ to allow for warping of the cross-section, E is the elastic modulus and v the Poisson's ratio. In other situations such as in cellular plates or plates of sandwich construction, $\mathbf{D}_f$ and $\mathbf{D}_s$ can be readily obtained using standard procedures discussed in [5].

The total potential energy may now be written as

$$\pi = \frac{1}{2} \int_A ([\chi]^T \mathbf{D}_f \chi + [\phi]^T \mathbf{D}_s \phi) \, dA - \int_A qw \, dA.$$

We now consider briefly two special features of thick plate theory. It is well known that in thin plate theory at any given point only two independent boundary conditions can be applied in terms of either displacements or tractions. This causes the introduction of such artifices as the "Kirchhoff Shear", and can lead to apparently inconsistent boundary conditions. In the analysis of thick plates with shear deformation, three independent boundary conditions are introduced and on occasion the analyst accustomed to thin plate theory may find these confusing. This arises in particular when displacements on edges have to be prescribed. In thin plate theory specification of deflection w immediately fixes the rotations about directions normal to the boundary θ_s as these are simply equal to $\partial w/\partial s$, where s is the distance along the restrained edge. Now in thick plate theory this rotation is independent and θ_s should be independently constrained. Much obviously depends on the real physical constraint applied and the solution with or without such a constraint can be physically justified depending on the nature of the support. For thin plates at least it is necessary to specify this rotation as zero for both simply supported and clamped plates for best agreement with conventional theory.

In thin plate theory "point loads" cause theoretically infinitely large bending moments at the point of application but the displacements under such loads are still finite. If shear deformation is considered, point loads will now cause also theoretically infinite displacements. Neither position is satisfactory from the engineer's viewpoint (although naturally the finite element approximation will give finite though large values). It is therefore suggested that wherever so-called point loads are encountered these should be spread over a "reasonable" finite area.

6.2 Shape functions and their Cartesian derivatives

In the parabolic isoparametric plate bending element, the geometry of the element is defined in an identical manner to that for the plane stress/strain element described in the previous chapter. Thus the geometry is defined by the expression

$$\begin{bmatrix} x \\ y \end{bmatrix} = \sum_{i=1}^{8} \mathbf{N}_i \begin{bmatrix} x_i \\ y_i \end{bmatrix} \tag{6.8}$$

where the shape function matrix associated with node i is

$$\mathbf{N}_i = N_i \mathbf{I}_3,$$

$\mathbf{I}_3$ is the 3×3 identity matrix and N_i is the shape function associated with node i as defined in the previous chapter. The displacement variation over the element is defined in terms of the nodal displacement components by the expression

$$\begin{bmatrix} w \\ \theta_x \\ \theta_y \end{bmatrix} = \sum_{i=1}^{8} \mathbf{N}_i \boldsymbol{\delta}_i, \tag{6.9}$$

where $\boldsymbol{\delta}_i = [w_i, \theta_{xi}, \theta_{yi}]^T$ is the vector of displacements at node i.

As with the plane stress/strain element, the same interpolation functions are used to describe geometry and the displacement variation; hence the term "isoparametric". The subroutine which calculates the shape functions and their derivatives is called SFR2 and has been described in the previous chapter in Section 5.3.

The shape function derivatives may again be calculated in the usual way using the chain rule.

$$\frac{\partial N_i}{\partial x} = \frac{\partial N_i}{\partial \xi} \cdot \frac{\partial \xi}{\partial x} + \frac{\partial N_i}{\partial \eta} \cdot \frac{\partial \eta}{\partial x}$$

$$\frac{\partial N_i}{\partial y} = \frac{\partial N_i}{\partial \xi} \cdot \frac{\partial \xi}{\partial y} + \frac{\partial N_i}{\partial \eta} \cdot \frac{\partial \eta}{\partial y}. \tag{6.10}$$

Derivatives $\partial \xi / \partial x$ etc. may be calculated from the inverse of the Jacobian matrix $[\mathbf{J}]^{-1}$. These values will be required when we calculate the strain matrix $\mathbf{B}$. Also, the determinant of the Jacobian matrix will be required for any integrations carried out over the element since an element of area is given as

$$\mathrm{d}x\,\mathrm{d}y = \det \mathbf{J}\,\mathrm{d}\xi\,\mathrm{d}\eta.$$

The subroutine which calculates the Cartesian shape function derivatives and the determination of the Jacobian matrix is called JACOB2 and has been described in the previous Chapter in Section 5.5. This subroutine calculates those quantities at the sampling points and also calculates the coordinates of the sampling points.

6.3 Strain matrix—B

The generalised strain/displacement relationship for the plate bending element may be written as

$$\boldsymbol{\varepsilon} = \sum_{i=1}^{8} \mathbf{B}_i \boldsymbol{\delta}_i, \tag{6.11}$$

where

$$[\boldsymbol{\varepsilon}]^{\mathrm{T}} = [[\boldsymbol{\chi}]^{\mathrm{T}}, [\boldsymbol{\phi}]^{\mathrm{T}}],$$

and

$$\mathbf{B}_i = \begin{bmatrix} \mathbf{B}_{fi} \\ \mathbf{B}_{si} \end{bmatrix} = \begin{bmatrix} 0 & -\dfrac{\partial N_i}{\partial x} & 0 \\[2ex] 0 & 0 & -\dfrac{\partial N_i}{\partial y} \\[2ex] 0 & -\dfrac{\partial N_i}{\partial y} & -\dfrac{\partial N_i}{\partial x} \\[1ex] \hline \dfrac{\partial N_i}{\partial x} & -N_i & 0 \\[2ex] \dfrac{\partial N_i}{\partial y} & 0 & -N_i \end{bmatrix} \tag{6.12}$$

where $\mathbf{B}_{fi}$ is the strain matrix associated with bending deformation χ and $\mathbf{B}_{si}$ is the strain matrix associated with shear deformation ϕ.

6.4 Subroutine BMATPB

This subroutine calculates the strain matrix **B** for the plate bending element using the shape functions and their Cartesian derivatives. The variable names used are identical to those used in the previous chapter.

Input/Output diagram

```
                INPUT                                           OUTPUT

                ⎡CARTD(NDIME,NNODE)
C(WORK)         ⎣SHAPE(NNODE)          →BMATPB→  BMATX(NSTRE,NEVAB) ⎤ C(WORK)

                ⎡NNODE
C(CONTRO)       ⎢NSTRE
                ⎣NEVAB
```

Annotated FORTRAN listing

```
        SUBROUTINE BMATPB
C
C***  CALCULATES STRAIN MATRIX B
C     FOR PLATE BENDING ELEMENT
C
```

┌─────────────────────────┐
│   COMMON BLOCKS   │
└─────────────────────────┘

```
    ┌──► DO 10 ISTRE=1,NSTRE
    │ ┌► DO 10 IEVAB=1,NEVAB                        ⎫
    │ │  BMATX(ISTRE,IEVAB)=0.0                      ⎬ 1*
    └─┴─10 CONTINUE                                  ⎭
         JGASH=0                                       2*
    ┌──► DO 20 INODE=1,NNODE                           3*
    │    IGASH=JGASH+1
    │    BMATX(4,IGASH)=CARTD(1,INODE)
    │    BMATX(5,IGASH)=CARTD(2,INODE)
    │    IGASH=IGASH+1                                ⎫
    │    JGASH=IGASH+1                                │
    │    BMATX(1,IGASH)=-CARTD(1,INODE)              │
    │    BMATX(3,IGASH)=-CARTD(2,INODE)              ⎬ 4*
    │    BMATX(4,IGASH)=-SHAPE(INODE)                │
    │    BMATX(2,JGASH)=-CARTD(2,INODE)              │
    │    BMATX(3,JGASH)=-CARTD(1,INODE)              │
    │    BMATX(5,JGASH)=-SHAPE(INODE)                ⎭
    └─20 CONTINUE
         RETURN
         END
```

1* Zero the **B** matrix.

2* Initialise JGASH which is a counter used to locate positions in the **B** matrix.

3* Loop for each node of the element.

4* Complete the $\mathbf{B}_i$ matrix where $\mathbf{B} = [\mathbf{B}_1, \mathbf{B}_2, \ldots, \mathbf{B}_8]$

$$\mathbf{B}_i = \begin{bmatrix} 0 & -\dfrac{\partial N_i}{\partial x} & 0 \\[2ex] 0 & 0 & -\dfrac{\partial N_i}{\partial y} \\[2ex] 0 & -\dfrac{\partial N_i}{\partial y} & -\dfrac{\partial N_i}{\partial x} \\[2ex] \dfrac{\partial N_i}{\partial x} & -N_i & 0 \\[2ex] \dfrac{\partial N_i}{\partial y} & 0 & -N_i \end{bmatrix}.$$

6.5 The matrix of elastic rigidities D

The generalised stress/strain relationships are identical to those given in (6.6). It should be noted that this element can be used for sandwich plates or cellular plates provided the **D** matrix is known [5].

6.6 Subroutine MODPB

Subroutine MODPB calculates the coefficients of the matrix of elastic rigidities **D** for the current element using the element material properties.

Input/Output diagram

```
            INPUT                                        OUTPUT

C(CONTRO)[NSTRE                    →| MODPB |→ DMATX(NSTRE,NSTRE)]C(WORK)
C(LGDATA)[PROPS(NMATS,NPROP)
A        [LPROP
```

Annotated FORTRAN listing

```
          SUBROUTINE MODPB(LPROP)
   C
   C*** CALCULATES MATRIX OF ELASTIC RIGIDITIES
   C    FOR PLATE BENDING ELEMENT
   C

      ┌──────────────────────────────┐
      │   COMMON  BLOCKS              │
      └──────────────────────────────┘

    ┌──►DO 10 ISTRE=1,NSTRE
    │ ┌►DO 10 JSTRE=1,NSTRE
    │ │  DMATX(ISTRE,JSTRE)=0.0          }1*
    └─10 CONTINUE
         YOUNG=PROPS(LPROP,1)
         POISS=PROPS(LPROP,2)            }2*
         THICK=PROPS(LPROP,3)
         DMATX(1,1)=YOUNG*THICK*THICK*THICK
        ./(12.0*(1.0-POISS*POISS))
         DMATX(1,2)=POISS*DMATX(1,1)
         DMATX(2,2)=DMATX(1,1)
         DMATX(2,1)=DMATX(1,2)           }3*
         DMATX(3,3)=(1.0-POISS)*DMATX(1,1)/2.0
         DMATX(4,4)=YOUNG*THICK/(2.4*(1.0+POISS))
         DMATX(5,5)=DMATX(4,4)
         RETURN
         END
```

1* Zero the **D** matrix.
2* Determine Young's modulus, Poisson's ratio and the thickness for the element currently under consideration.

3* Evaluate the **D** matrix for the plate element.

where

$$
\mathbf{D} = \begin{bmatrix}
\dfrac{Et^3}{12(1-v^2)} & \dfrac{vEt^3}{12(1-v^2)} & 0 & 0 & 0 \\[2ex]
\dfrac{vEt^3}{12(1-v^2)} & \dfrac{Et^3}{12(1-v^2)} & 0 & 0 & 0 \\[2ex]
0 & 0 & \dfrac{(1-v)}{2}\dfrac{Et^3}{12(1-v^2)} & 0 & 0 \\[2ex]
0 & 0 & 0 & \dfrac{Et}{2\cdot4(1+v)} & 0 \\[2ex]
0 & 0 & 0 & 0 & \dfrac{Et}{2\cdot4(1+v)}
\end{bmatrix}
$$

6.7 Stress matrix DB

Since the strain/displacement relationship in the finite element approximation may be written as

$$\boldsymbol{\varepsilon} = [\mathbf{B}_1, \ldots, \mathbf{B}_8]\boldsymbol{\delta}^e, \tag{6.13}$$

the stress at any point within the element can be expressed as

$$\boldsymbol{\sigma} = \mathbf{DB}\boldsymbol{\delta}^e = \mathbf{S}\boldsymbol{\delta}^e. \tag{6.14}$$

The subroutine DBE which multiplies matrix **D** by matrix **B** is identical to that used for the two previous applications and since it is described in Section 4.10.6 it will not be repeated here. Note that the product **DB = S** is stored in DBMAT(ISTRE,IEVAB) where ISTRE ranges over the number of in-plane stress components (=5 for plates) and IEVAB ranges over the element variables.

We should note here that a certain economy could have been obtained in the multiplication of **D** by **B** which avoids unnecessary multiplication. This could have been achieved by use of the following expression instead of (6.14).

$$\boldsymbol{\sigma} = \begin{bmatrix} \boldsymbol{\sigma}_f \\ \boldsymbol{\sigma}_s \end{bmatrix} = \begin{bmatrix} \mathbf{D}_f\mathbf{B}_f \\ \mathbf{D}_s\mathbf{B}_s \end{bmatrix}\boldsymbol{\delta}^e. \tag{6.15}$$

This expression was not used in order that a modular subroutine, namely DBE, could be used for all applications. The reader should appreciate that one must expect some loss in economy if one requires greater modularity of subroutines.

6.8 The element stiffness matrix

We now have all of the information necessary to calculate the element stiffness matrix $\mathbf{K}^e$.

A typical submatrix of the element stiffness matrix linking nodes p and q may be written as

$$\mathbf{K}_{pq} = \int\int [\mathbf{B}_p]^T \mathbf{D} \mathbf{B}_q \, dx \, dy, \qquad (6.16)$$

where

$$\mathbf{D} = \begin{bmatrix} \mathbf{D}_f & 0 \\ 0 & \mathbf{D}_s \end{bmatrix}.$$

It is important to note a certain economy which has *not* been incorporated into the program. The flexural contribution to the submatrix may be defined by the expression

$$\mathbf{K}_{pqf} = \int\int [\mathbf{B}_{pf}]^T \mathbf{D}_f \mathbf{B}_{qf} \, dx \, dy. \qquad (6.17)$$

Similarly, the shear contribution to the submatrix may be defined by the expression

$$\mathbf{K}_{pqs} = \int\int [\mathbf{B}_{ps}]^T \mathbf{D}_s \mathbf{B}_{qs} \, dx \, dy, \qquad (6.18)$$

by evaluating the stiffness matrix in this manner, many unnecessary matrix multiplications may be avoided as

$$\mathbf{K}_{pq} = \mathbf{K}_{pqf} + \mathbf{K}_{pqs}.$$

The integration of the stiffness coefficients is carried out numerically using a 2×2 Gauss rule as described in an earlier section. Experience [6] has shown this to be the optimum integration rule as it eliminates spurious shear strain energy effects when the plate is thin. Some further theoretical aspects of this element are described elseshere in [5].

6.9 Subroutine STIFPB

This routine which calculates the element stiffness and stress matrices also contains all the features of routines STIFB and STIFPS which calculate the stiffness matrix, etc. for the beam and plane stress/strain applications respectively. The variables used here are described in Sections 4.10 and 5.13 and will not be repeated here.

Input/Output diagram

INPUT   OUTPUT

$$\rightarrow \boxed{\text{STIFPB}} \rightarrow$$

C(CONTRO) $\begin{bmatrix} \text{NELEM} \\ \text{NDIME} \\ \text{NNODE} \\ \text{NEVAB} \\ \text{NGAUS} \\ \text{NSTRE} \end{bmatrix}$

ESTIF(NEVAB,NEVAB)]F(1)

SMATX(NSTRE,NEVAB,NGASP)
GPCOD(NDIME,NGASP) $\Big]$F(3)

C(LGDATA) $\begin{bmatrix} \text{PROPS(NMATS,NPROP)} \\ \text{MATNO(NELEM)} \\ \text{LNODS(NELEM,NNODE)} \\ \text{COORD(NPOIN,NNODE)} \\ \text{WEIGP(NGAUS)} \\ \text{POSGP(NGAUS)} \end{bmatrix}$

C(WORK) $\begin{bmatrix} \text{BMATX(NSTRE,NEVAB)} \\ \text{DBMAT(NSTRE,NEVAB)} \end{bmatrix}$

A(JACOB2) [DJACB

Annotated FORTRAN listing

```
        SUBROUTINE STIFPB
C
C***  CALCULATES ELEMENT STIFFNESS MATRIX
C     FOR PLATE BENDING ELEMENT
C
      DIMENSION ESTIF(24,24)
```

```
┌─────────────────────────┐
│    COMMON  BLOCKS        │
└─────────────────────────┘
```

```
C
C*** LOOP OVER EACH ELEMENT
C
     ┌──►DO 70 IELEM=1,NELEM                    1 *
     │   LPROP=MATNO(IELEM)                      2 *
     │ C
     │ C*** EVALUATE THE COORDINATES OF THE
     │ C    ELEMENT NODAL POINTS
     │ C
     │ ┌──►DO 10 INODE=1,NNODE            ⎫
     │ │   LNODE=LNODS(IELEM,INODE)       ⎬
     │ ├──►DO 10 IDIME=1,NDIME            ⎬ 3 *
     │ │   ELCOD(IDIME,INODE)=COORD(LNODE,IDIME) ⎭
     │ └══10 CONTINUE
     │ C
     │ C*** INITIALIZE THE ELEMENT STIFFNESS MATRIX
     │ C
```

```
      DO 20 IEVAB=1,NEVAB
        DO 20 JEVAB=1,NEVAB                        } 4 *
          ESTIF(IEVAB,JEVAB)=0.0
   20 CONTINUE
C
C***    CALCULATE MATRIX OF ELASTIC RIGIDITIES
C
        CALL MODPB(LPROP)                          5 *
        KGASP=0                                    6 *
C
C***  ENTER LOOPS FOR NUMERICAL INTEGRATION
C
      DO 50 IGAUS=1,NGAUS                          7 *
        EXISP=POSGP(IGAUS)                         8 *
        DO 50 JGAUS=1,NGAUS                        9 *
        ETASP=POSGP(JGAUS)                        10 *
        KGASP=KGASP+1                             11 *
C
C***  EVALUATE THE SHAPE FUNCTIONS,
C     ELEMENTAL AREA,ETC.
C
        CALL SFR2(EXISP,ETASP)                    12 *
        CALL JACOB2(IELEM,DJACB,KGASP)            13 *
        DAREA=DJACB*WEIGP(IGAUS)*WEIGP(JGAUS)     14 *
C
C***  EVALUATE THE B AND DB MATRICES
C
        CALL BMATPB                               15 *
        CALL DBE                                  16 *
C
C***  CALCULATE THE ELEMENT STIFFNESS
C
      DO 30 IEVAB=1,NEVAB
        DO 30 JEVAB=IEVAB,NEVAB                    } 17 *
        DO 30 ISTRE=1,NSTRE
          ESTIF(IEVAB,JEVAB)=ESTIF(IEVAB,JEVAB)+
         .BMATX(ISTRE,IEVAB)*DBMAT(ISTRE,JEVAB)    } 18 *
         .*DAREA
   30 CONTINUE
C
C***  STORE THE COMPONENTS OF THE DB MATRIX
C     FOR THE ELEMENT
C
      DO 40 ISTRE=1,NSTRE
        DO 40 IEVAB=1,NEVAB                        }
        SMATX(ISTRE,IEVAB,KGASP)=                   19 *
         .DBMAT(ISTRE,IEVAB)
   40 CONTINUE
   50 CONTINUE
C                                                   20 *
C***  CONSTRUCT THE LOWER TRIANGLE
C     OF THE STIFFNESS MATRIX
C
```

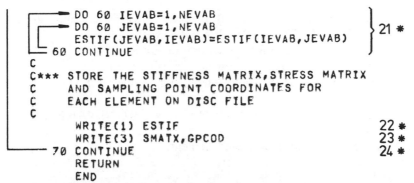

```
      ┌──── DO 60 IEVAB=1,NEVAB                              ⎫
      │ ┌── DO 60 JEVAB=1,NEVAB                              ⎪  21 *
      │ │   ESTIF(JEVAB,IEVAB)=ESTIF(IEVAB,JEVAB)            ⎬
      │ └─ 60 CONTINUE                                       ⎭
      │ C
      │ C*** STORE THE STIFFNESS MATRIX,STRESS MATRIX
      │ C    AND SAMPLING POINT COORDINATES FOR
      │ C    EACH ELEMENT ON DISC FILE
      │ C
      │      WRITE(1) ESTIF                                    22 *
      │      WRITE(3) SMATX,GPCOD                              23 *
      └─ 70 CONTINUE                                           24 *
             RETURN
             END
```

1* Loop over each element.

2* Identify the material property type of the element.

3* Store the coordinates of the nodal point of the element under consideration in the local array ELCOD(NDIME,NNODE).

4* Set the element stiffness array ESTIF(NEVAB,NEVAB) to zero.

5* Call subroutine MODPB(LPROP) which sets up the **D** matrix for the material.

6* Set index KGASP to zero. This counter will be used to indicate the Gauss point number of an element.

7* Enter ξ loop for numerical integration.

8* Determine ξ coordinate of current Gauss point.

9* Enter η loop for numerical integration.

10* Determine η coordinate of current Gauss point.

11* Increment KGASP by one to give the current Gauss point number.

12* Evaluate the shape functions N_i and the derivatives $\partial N_i/\partial \xi$ and $\partial N_i/\partial \eta$ at current sampling point.

13* Calculate the Jacobian matrix **J**, its determinant DJACB, its inverse $[\mathbf{J}]^{-1}$ and the Cartesian derivatives of the shape functions. This subroutine also calculates the Cartesian coordinates x, y of the Gauss points which are stored in array GPCOD.

14* Calculate the elemental area DAREA for numerical integration.

15* Call subroutine BMATPB which generates the **B** matrix for the current Gauss point.

16* Call subroutine DBE which forms the matrix product **DB** for the current Gauss point.

17* Enter two loops over the range NEVAB where NEVAB is the total number of degrees of freedom per element. These loops control the construction of the element stiffness matrix ESTIF(NEVAB,NEVAB).

18* Construct the element stiffness matrix. It should be noted that only the upper triangle is formed at this stage due to DO LOOP parameters in 17*.

19* Store the matrix product **DB** for each Gaussian sampling point in the array SMATX(NSTRE,NEVAB,KGASP).
20* Termination of loop for numerical integration.
21* Complete the element stiffness matrix using symmetry.
22* Write onto a disc file the element stiffness matrix ESTIF(NEVAB, NEVAB).
23* Also write the stress matrix SMATX(NSTRE,NEVAB,KGASP) to another file.
24* Terminate DO loop over each element.

6.10 Calculation of element stress resultants

Since the Gaussian sampling points within each element have already been defined it is convenient to output the stress resultants at these points.

The stress resultants are readily calculated from the displacements as described in Section 6.7 and for this purpose the matrix product **DB** has already been evaluated at each element Gaussian point and stored on disc file as the stress matrix SMATX.

6.11 Subroutine STREPB

This subroutine is called after the displacements have been evaluated and calculates the stress resultants at the Gauss points. The variables employed are identical to those used in Section 5.15 for the plane stress/strain appli-application.

Input/Output diagram

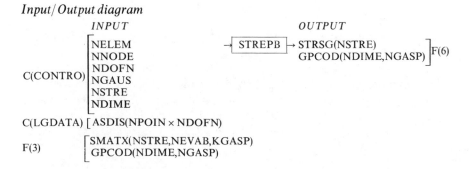

Annotated FORTRAN listing

```
          SUBROUTINE STREPB
C
C***   CALCULATES STRESS RESULTANTS AT GAUSS
C      POINTS FOR PLATE BENDING ELEMENT
C
          DIMENSION ELDIS(3,8),STRSG(5)
```

┌─────────────────────────┐
│ COMMON BLOCKS   │
└─────────────────────────┘

```
          WRITE(6,900)                                    1*
          WRITE(6,905)
C
C***   LOOP OVER EACH ELEMENT
C
          DO 40 IELEM=1,NELEM                             2*
C
C***   READ THE STRESS MATRIX , SAMPLING POINT
C      COORDINATES FOR THE ELEMENT
C
          READ(3) SMATX,GPCOD                             3*
          WRITE(6,910) IELEM                              4*
C
C***   IDENTIFY THE DISPLACEMENTS OF THE
C      ELEMENT NODAL POINTS
C
          DO 10 INODE=1,NNODE
          LNODE=LNODS(IELEM,INODE)
          NPOSN=(LNODE-1)*NDOFN
          DO 10 IDOFN=1,NDOFN                             5*
          NPOSN=NPOSN+1
          ELDIS(IDOFN,INODE)=ASDIS(NPOSN)
     10   CONTINUE
          KGASP=0                                         6*
C
C***   ENTER LOOPS OVER EACH SAMPLING POINT
C
          DO 30 IGAUS=1,NGAUS                             7*
          DO 30 JGAUS=1,NGAUS
          KGASP=KGASP+1                                   8*
          DO 20 ISTRE=1,NSTRE
          STRSG(ISTRE)=0.0
          KGASH=0
C
C***   COMPUTE THE STRESS RESULTANTS
C                                                         9*
          DO 20 INODE=1,NNODE
          DO 20 IDOFN=1,NDOFN
          KGASH=KGASH+1
          STRSG(ISTRE)=STRSG(ISTRE)+
         .SMATX(ISTRE,KGASH,KGASP)*ELDIS(IDOFN,INODE)
     20   CONTINUE
```

```
      C
      C*** OUTPUT THE STRESS RESULTANTS
      C
            WRITE(6,915) KGASP,                                    10*
           .(GPCOD(IDIME,KGASP),IDIME=1,NDIME),
           .(STRSG(ISTRE),ISTRE=1,NSTRE)
       30 CONTINUE                                                 11*
       40 CONTINUE                                                 12*
      900 FORMAT(/,10X,8HSTRESSES,/)
      905 FORMAT(1H0,4HG,P,,2X,8HX=COORD,,2X,
           .8HY=COORD,,3X,8HX=MOMENT,4X,8HY=MOMENT,
           .3X,9HXY=MOMENT,2X,10HXZ=S,FORCE,2X,
           .10HYZ=S,FORCE)
      910 FORMAT(/,5X,12HELEMENT NO,=,I5)
      915 FORMAT(I5,2F10,4,5E12,5)
          RETURN
          END
```

1* Write title for stress resultants.
2* Loop over each element.
3* Read from file 3 the matrix product **DB** and Gauss point coordinates for each Gauss point of the current element.
4* Write heading naming current element.
5* Store the displacements of the nodal points associated with the current element in the local array ELDIS(NDIME,NNODE).
6* Set KGASP = 0. This index is used to identify the Gauss point number for use in the SMATX array. Here, as soon as an element has been completely processed KGASP is reset to zero.
7* Enter loops for processing each element Gauss point. It should be noted that no numerical integration is performed in this instance.
8* Increment KGASP by one to give the location of the current Gauss point for use in the SMATX array.
9* Evaluate the stress resultants using the expression

$$\sigma = \mathbf{DB}\,\delta^e.$$

10* Output the stress resultants and the Gauss point Cartesian coordinates.
11* End of Gauss loop.
12* End of element loop.

References

1. Zienkiewicz, O. C., "The Finite Element Method in Engineering Science". 2nd edition, McGraw-Hill, London 1971.
2. Zienkiewicz, O. C., and Cheung, Y. K., The finite element method for analysis of elastic isotropic and orthotropic slabs. *Proc. Inst. Civ. Eng.* **28**, 471–488, 1964.

3. Cheung, Y. K., King, I. P., and Zienkiewicz, O. C. Slab bridges with arbitrary shape and support conditions: a general method of analysis based on finite elements. *Proc. Inst. Civ. Eng.* **40**, 9–36, 1968.
4. Mindlin, R. D., Influence of rotatory inertia and shear on flexural motions of isotropic elastic plates. *J. Appl. Mech.*, **18**, 31–38, 1951.
5. Hinton, E., Razzaque, A., Zienkiewicz, O. C., and Davies, J. D., A simple finite element solution for plates of homogeneous, sandwich and cellular construction. *Proc. Inst. Civ. Eng.* **59**, Part 2. 43–65, 1975.
6. Zienkiewicz, O. C., Taylor, R. L., and Too, J. M., Reduced integration technique in general analysis of plates and shells. *Int. J. Num. Meth. Eng.* **3**, 275–290, 1971.

7

Equivalent Nodal
Representation of Loads

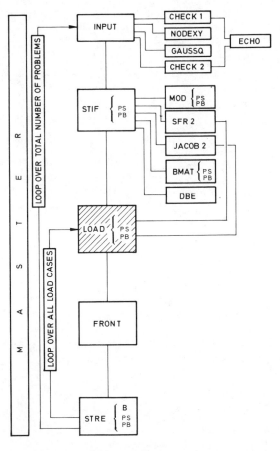

7.1 Introduction

In the finite element analysis of structures by the displacement method, the only permissible form of loading, other than initial stressing, is by the prescription of concentrated loads at the nodal points. Consequently forms of loading such as gravity action, pressures assigned to element surfaces, must be reduced to equivalent nodal forces before solution can proceed. In any finite element program based on the isoparametric element principle the calculation of equivalent nodal forces is not a process that can be performed manually since area or volume integrations over arbitrarily shaped regions are generally involved. Hence the equivalent nodal forces, due to pressures, etc., cannot be calculated for direct input and the inclusion of subroutines to perform this task is necessary. The object of this Chapter is to describe and formulate such subroutines for the various loadings encountered in plane stress/strain situations and plate bending problems.

As an introduction to the principles involved we will consider the case of a point load acting on a parabolic isoparametric beam element when the point of application is not coincident with a nodal point as shown in Fig. 7.1(a). Suppose the shape functions corresponding to the three nodal points are N_1, N_2 and N_3. Now let a virtual displacement w^*, in the z direction occur at each node in turn and let us denote the lateral nodal forces at each node which are equivalent to the applied load P as P_1, P_2 and P_3. When the virtual displacement is applied to node i, by the principle of virtual work we have

$$P_i w^* = P . N_i(\xi_P) w^*, \qquad (7.1)$$

where N_i is evaluated for the value of ξ at which the point load acts. Since this expression is to hold for an arbitrary displacement, w^*, then

$$P_i = P . N_i(\xi_P). \qquad (7.2)$$

This expression allows the equivalent nodal forces P_i, to be determined by setting $i = 1$ to 3 in turn.

(We note in passing that with this beam, applied lateral point forces produce only applied lateral nodal point forces with no nodal couples. This is because the lateral displacement w is independent of the rotation of the normal since $\theta = (\partial w/\partial x) + \phi$. With the conventional "cubic displacement" thin beam the lateral displacement w is not independent of the rotation of the normal since $\theta = \partial w/\partial x$ and therefore a lateral point force produces lateral nodal forces *and* couples.)

We turn now to the more general case of a point load applied to the edge of a parabolic isoparametric plane stress element shown in Fig. 7.1(b).

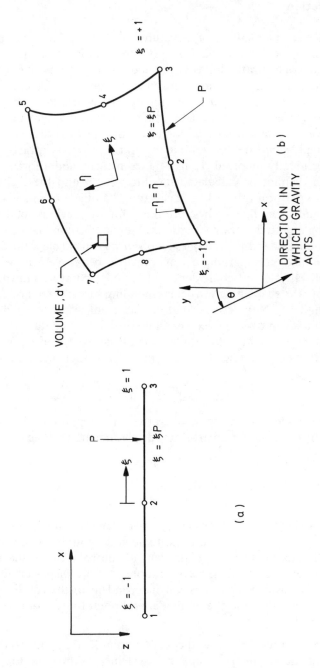

FIG. 7.1. Application of point loads to one- and two-dimensional parabolic isoparametric elements.

The principle we have just discussed can again be invoked. Here virtual displacements are applied in the x and y directions to each node in turn. Thus for node i we have for virtual displacement components u^* and v^* respectively

$$P_{xi}u^* = P_x N_i(\xi_P, \bar{\eta})u^*$$
$$P_{yi}v^* = P_y N_i(\xi_P, \bar{\eta})v^*, \tag{7.3}$$

where P_x and P_y are the components of the load P acting in the x and y directions respectively and ξ_P is the ξ coordinate of the point and $\bar{\eta}$ is the constant value of η at the edge in question i.e. $\bar{\eta} = -1$ or $+1$. Once again invoking the arbitrary nature of the virtual displacement

$$\begin{bmatrix} P_{xi} \\ P_{yi} \end{bmatrix} = N_i(\xi_P, \bar{\eta}) \begin{bmatrix} P_x \\ P_y \end{bmatrix}. \tag{7.4}$$

Thus again the equivalent nodal forces can be calculated as soon as the load components P_x and P_y and its point of application, as defined by $(\xi_P, \bar{\eta})$, are specified. It should be noted that $N_i(\xi_P, \bar{\eta}) = 0$ and the corresponding forces P_{xi}, P_{yi} are zero, except for nodes along the loaded edge.

The most convenient way of specifying the point of application would be by prescribing the x and y coordinates of this point. If the nodal point coordinates are (x_i, y_i) then by the isoparametric concept the coordinates of any point along the element are given by

$$\begin{bmatrix} x \\ y \end{bmatrix} = \sum_{i=1}^{3} N_i \begin{bmatrix} x_i \\ y_i \end{bmatrix}. \tag{7.5}$$

If the x and y coordinates of the point of application are specified as x_p and y_p respectively then

$$\begin{bmatrix} x_P \\ y_P \end{bmatrix} = \sum_{i=1}^{3} N_i(\xi_P, \bar{\eta}) \begin{bmatrix} x_i \\ y_i \end{bmatrix}. \tag{7.6}$$

The shape functions N_i are, in this case, quadratic expressions in ξ and η and therefore (7.6) is quadratic in ξ and can be solved to yield ξ_P. The equivalent nodal forces can then be determined from (7.4). (This, of course, implies that the point of application of the load is known to coincide with a particular mesh line).

We will now consider in turn the various forms of loading which may occur in the structural applications considered in this text. For the beam element the loads are formulated at the same time as the stiffnesses, the process having already been described in Chapter 4.

7.2 Gravity loading for plane stress/strain situations

Gravity forces are equivalent to a body force/unit volume acting within the solid in the direction of the gravity axis. In the following the gravity axis need not be coincident with either of the coordinate axes and consequently gravity force components may act in both the x and y directions. The direction in which gravity acts will be defined by specifying the angle which the gravity axis makes with the y axis.

To obtain the equivalent nodal forces, the virtual work concept introduced in the previous section will again be employed. Figure 7.1(b) illustrates an isoparametric element subjected to gravity forces acting at an angle θ measured anti-clockwise from the positive y axis as shown. If g is the gravitational acceleration and the material mass density is ρ, then the gravity force dG acting on an elemental volume dV is

$$dG = \rho g\, dV, \tag{7.7}$$

which acts in the direction of the gravity axis. The components acting in the x and y directions respectively are

$$dG_x = \rho g\, dV \sin\theta$$
$$dG_y = -\rho g\, dV \cos\theta. \tag{7.8}$$

As in the previous section we now apply virtual displacements u^* and v^* in the x and y directions respectively at each node in turn. Applying the principle of virtual work then results in the following expressions for the equivalent nodal forces P_{xi} and P_{yi}, where i ranges from 1 to 8 in this case

$$P_{xi}u^* = \int_{V_e} N_i u^* \rho g \sin\theta\, dV,$$

$$P_{yi}v^* = -\int_{V_e} N_i v^* \rho g \cos\theta\, dV, \tag{7.9}$$

where N_i are the shape functions and integration is taken over the volume of the element. Or again since (7.9) holds for all values of u^* and v^*

$$\begin{bmatrix} P_{xi} \\ P_{yi} \end{bmatrix} = \int_{V_e} N_i \rho g \begin{bmatrix} \sin\theta \\ -\cos\theta \end{bmatrix} dV. \tag{7.10}$$

The equivalent nodal forces due to a gravitational acceleration g acting on a material of mass density ρ can be determined from this expression. In performing the integration in (7.10) it is again essential to resort to a Gaussian numerical integration technique, resulting in (7.10) being replaced, for practical purposes by

$$\begin{bmatrix} P_{xi} \\ P_{yi} \end{bmatrix} = \sum_{n=1}^{\text{NGAUS}} \sum_{m=1}^{\text{NGAUS}} \rho g t \begin{bmatrix} \sin \theta \\ -\cos \theta \end{bmatrix} N_i(\xi_n, \eta_m) W_n W_m \det \mathbf{J}, \qquad (7.11)$$

where t is the element thickness, $\mathbf{J}$ is the Jacobian matrix defined in Chapter 5 and W_n, W_m are the Gaussian weighting functions also discussed in Chapter 5.

In equation solution by the frontal approach, an element is processed as soon as all the information pertaining to it has been assimilated by the solution routine. Immediately the stiffness and load terms of an element have been accepted, assembly and elimination follows directly. Consequently it is convenient to store the nodal forces associated with each element separately, and not assemble the total force at a node by summation of the contributions of all attached elements. Thus the equivalent nodal forces are stored in the following array ELOAD(IELEM,IEVAB), where IELEM indicates the element under consideration and IEVAB ranges over the total number of degrees of freedom, NEVAB, of the element. We are now in a position to formulate the portion of the loading subroutine for plane stress/strain problems which evaluates the equivalent nodal forces due to gravity. This section will be termed the GRAVITY SECTION and will be later incorporated into the finished subroutine LOADPS.

All the variables employed in this section have previously been defined in Chapters 3–6 and therefore we can proceed with the programming immediately. Explanatory notes will again be provided.

Input/Output diagram

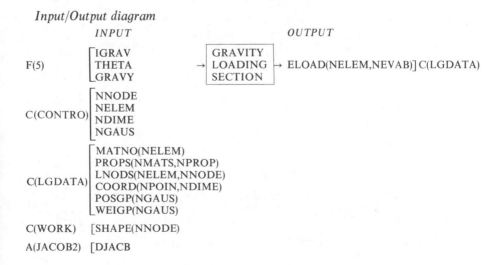

Annotated FORTRAN listing

```
C
C*** GRAVITY LOADING SECTION
C
C
C*** READ GRAVITY ANGLE AND GRAVITATIONAL
C    CONSTANT
C
      READ(5,920) THETA,GRAVY
  920 FORMAT(2F10.3)                                        ⎫
      WRITE(6,925) THETA,GRAVY                              ⎬  1*
  925 FORMAT(1H0,16H GRAVITY ANGLE =,F10.3,                 ⎭
     . 19H GRAVITY CONSTANT =,F10.3)
      THETA=THETA/57.295779514                                 2*
C
C*** LOOP OVER EACH ELEMENT
C
      DO 90 IELEM=1,NELEM                                      3*
C
C*** SET UP PRELIMINARY CONSTANTS
C
      LPROP=MATNO(IELEM)                                       4*
      THICK=PROPS(LPROP,3)                                  ⎫
      DENSE=PROPS(LPROP,4)                                  ⎬  5*
      IF(DENSE.EQ.0.0) GO TO 90                                6*
      GXCOM=DENSE*GRAVY*SIN(THETA)                          ⎫
      GYCOM=-DENSE*GRAVY*COS(THETA)                         ⎬  7*
C
C*** COMPUTE COORDINATES OF THE ELEMENT
C    NODAL POINTS
C
      DO 60 INODE=1,NNODE                                   ⎫
      LNODE=LNODS(IELEM,INODE)                              ⎬  8*
      DO 60 IDIME=1,NDIME                                   ⎭
   60 ELCOD(IDIME,INODE)=COORD(LNODE,IDIME)
C
C*** ENTER LOOPS FOR AREA NUMERICAL INTEGRATION
C
      DO 80 IGAUS=1,NGAUS                                   ⎫  9*
      DO 80 JGAUS=1,NGAUS                                   ⎭
      EXISP=POSGP(IGAUS)                                    ⎫ 10*
      ETASP=POSGP(JGAUS)                                    ⎭
C
C*** COMPUTE THE SHAPE FUNCTIONS AT THE SAMPLING
C    POINTS AND ELEMENTAL VOLUME
C
      CALL SFR2(EXISP,ETASP)                                   11*
      KGASP=1                                                  12*
      CALL JACOB2(IELEM,DJACB,KGASP)                           13*
      DVOLU=DJACB*WEIGP(IGAUS)*WEIGP(JGAUS)                 ⎫ 14*
      IF(THICK.NE.0.0) DVOLU=DVOLU*THICK                    ⎭
C
C*** CALCULATE LOADS AND ASSOCIATE WITH
```

```
C     ELEMENT NODAL POINTS
C
      DO 70 INODE=1,NNODE                                    15*
      NGASH=(INODE-1)*NDOFN+1
      MGASH=(INODE-1)*NDOFN+2
      ELOAD(IELEM,NGASH)=FLOAD(IELEM,NGASH)+
    . GXCOM*SHAPE(INODE)*DVOLU                               16*
   70 ELOAD(IELEM,MGASH)=ELOAD(IELEM,MGASH)+
    . GYCOM*SHAPE(INODE)*DVOLU
   80 CONTINUE
   90 CONTINUE                                               17*
```

1* Read and write the angle (in degrees) which the gravity axis makes with the y axis (Fig. 7.1b) and the gravitational acceleration.

2* Convert the gravity angle to radians.

3* Loop over each element.

4* Identify the material property type of the element.

5* Set THICK and DENSE equal to the element thickness and material density respectively.

6* If the material density is zero avoid the calculation of gravity loads for this element.

7* Set up the constant terms in the forces of (7.8).

8* Store the coordinates of the nodal points of the element under consideration in the local array ELCOD(IDIME,INODE).

9* Enter the loops for the numerical integration required in (7.11).

10* Set up the coordinates ξ_n, η_m of the Gaussian sampling points for use in (7.11).

11* Calculate the values of the shape functions and their derivatives corresponding to ξ_n and η_m i.e. N_i, $\partial N_i/\partial \xi$, $\partial N_i/\partial \eta$.

12* Set KGASP = 1 merely to avoid overstepping the bounds of array GPCOD in subroutine JACOB2. Array GPCOD is not used in the present calculation.

13* Calculate the Jacobian matrix $\mathbf{J}$, its determinant, DJACB, its inverse $[\mathbf{J}]^{-1}$ and the Cartesian derivatives of the shape functions.

14* Calculate the equivalent of the elemental volume, dV, for numerical integration, i.e. DVOLU = $W_n W_m$ det $\mathbf{J} . t$.
It should be noted that if the element thickness is not specified it is automatically assumed to be unity.

15* Loop over the element nodal connections.

16* Calculate the correct positions in the ELOAD array for inserting the equivalent nodal forces, e.g. for the Kth nodal connection number, the x and y components will be respectively stored in ELOAD(IELEM, ($K - 1$)*2 + 1) and ELOAD(IELEM,($K - 1$)*2 + 2). Then calculate the equivalent nodal forces and store in the correct locations.

17* Termination of DO LOOPS for numerical integration and total
number of elements respectively.

7.3 Normal and tangential loads/unit length applied to element edges for plane stress/strain situations

Any element edge will be allowed to have a distributed loading per unit
length in a normal and tangential direction prescribed to it as shown in
Fig. 7.2. These distributed forces need not be constant but can vary (independently) along the element edge. Since parabolic isoparametric elements

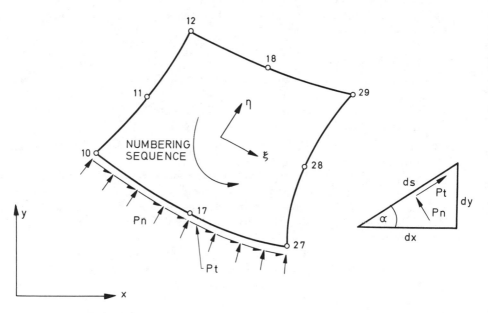

Fig. 7.2.   Normal and tangential loads/unit length applied to a parabolic isoparametric element.

are being employed, then at best a parabolic loading distribution can be
accommodated. The variation will be defined by prescribing the normal and
tangential values at the three nodal points forming the element edge to which
the loads are applied.

In order to be consistent with the order of listing of nodal connection numbers in the element topology definition, the three nodes forming the loaded edge must also be listed in an anticlockwise sequence with respect to the loaded element. Thus with regard to Fig. 7.2 the loaded element edge

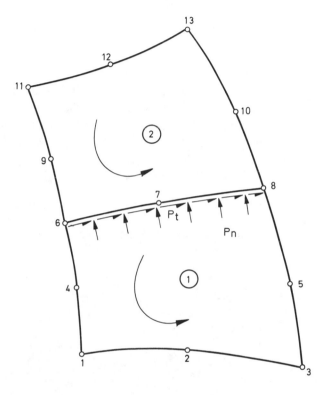

FIG. 7.3.   Sign convention for interfacial distributed loading.

would be defined by the nodal sequence

$$10 \qquad 17 \qquad 27$$

A pressure normal to a face is assumed to be positive if it acts in a direction *into* the element. A tangential load is assumed to be positive if it acts in an *anticlockwise* direction with respect to the loaded element. These definitions are necessary to avoid confusion when a distributed load acts on the interface between two elements as illustrated in Fig. 7.3.

Suppose that the three nodal values of normal and tangential distributed loads are $(p_n)_i$ and $(p_t)_i$ respectively, where i ranges from 1 to 3. Let the three shape functions defining a parabolic variation along the element edge be N_1, N_2 and N_3. These are, of course, the three shape functions employed in Chapter 4 for the one-dimensional beam. Then the distributed loading intensity at any point along the loaded edge is given by

$$\begin{bmatrix} p_n \\ p_t \end{bmatrix} = \sum_{i=1}^{3} N_i \begin{bmatrix} (p_n)_i \\ (p_t)_i \end{bmatrix}. \tag{7.12}$$

Again by applying virtual displacements u^* and v^* to each nodal point of the element in turn, we can determine the equivalent nodal forces. As a first step we must determine the components of the distributed loads acting in the x and y direction. Consider the forces acting on an incremental length dS of the loaded edge. The components of force acting in the x and y directions respectively are

$$dP_x = (p_t \, dS \cos \alpha - p_n \, dS \sin \alpha) = (p_t \, dx - p_n \, dy)$$

$$dP_y = (p_n \, dS \cos \alpha + p_t \, dS \sin \alpha) = (p_n \, dx + p_t \, dy). \tag{7.13}$$

Or since we wish to integrate along the element edge in terms of the curvilinear variable ξ we can write

$$dx = \frac{\partial x}{\partial \xi} \cdot d\xi; \qquad dy = \frac{\partial y}{\partial \xi} \cdot d\xi. \tag{7.14}$$

Using these in (7.13) we have,

$$dP_x = \left(p_t \frac{\partial x}{\partial \xi} - p_n \frac{\partial y}{\partial \xi} \right) d\xi$$

$$dP_y = \left(p_n \frac{\partial x}{\partial \xi} + p_t \frac{\partial y}{\partial \xi} \right) d\xi. \tag{7.15}$$

Then applying the principle of virtual work when u^* and v^* are applied to each node in turn we have the following expressions for the equivalent nodal forces

$$P_{xi} = \int_{S_e} N_i \left(p_t \frac{\partial x}{\partial \xi} - p_n \frac{\partial y}{\partial \xi} \right) d\xi$$

$$P_{yi} = \int_{S_e} N_i \left(p_n \frac{\partial x}{\partial \xi} + p_t \frac{\partial y}{\partial \xi} \right) d\xi, \tag{7.16}$$

where integration is taken along the loaded element edge, S_e. These expressions are in fact particular forms of the general expression (1.35). Again Gaussian numerical integration will be employed. We are now in a position to formulate the section of the loading subroutine which performs this function for plane situations. This portion will be termed the DISTRIBUTED EDGE LOADS SECTION and will later be incorporated into the completed subroutine LOADPS.

If a distributed load acts on the interface between two elements the required information can be input in two ways. With regard to Fig. 7.3 the loads can be considered to be acting on *either* element 1 *or* element 2.

(i) If the loading is presumed to act on element 1 the nodal specification of the element edge is

8   7   6

and will be stored in the array NOPRS(IODEG) where the index IODEG ranges from 1 to NODEG. The variable NODEG is the number of nodes defining the element edge and in this case is 3.

The values of the normal and tangential load intensities at these three nodal points will be input via the array

PRESS(IODEG,IDOFN)

with IODEG ranging from 1 to NODEG. The index IDOFN denotes the normal and tangential load components and consequently ranges from 1 to NDOFN which has a value of 2 for plane problems. Thus, for example, PRESS(3,2) contains the value of the tangential load intensity at node number 6 and if it acts in the direction indicated will have a negative value. Similarly for the direction indicated the normal pressure components will also be negative.

(ii) Alternatively the loading can be assumed to act on element 2. In this case the nodal specification of the element edge is

6   7   8

and the normal and tangential load intensities/unit length will have positive values if they act in the directions indicated. Thus for each element edge which is loaded by distributed loading the following information must be input.

NEASS,(NOPRS(IODEG),IODEG = 1,NODEG),
((PRESS(IODEG,IDOFN),IODEG = 1,NODEG),IDOFN = 1,
NDOFN)

where NEASS is the number of the element with which the edge loading is associated. The total number of loaded edges is defined by the variable NEDGE. The distributed edge loads section of the subroutine can now be listed.

Input/Output diagram

INPUT   OUTPUT

F(5)
$$\begin{bmatrix} \text{IEDGE} \\ \text{NEDGE} \\ \text{NOPRS(NODEG)} \\ \text{PRESS(NODEG,NDOFN)} \end{bmatrix} \rightarrow \boxed{\begin{array}{l}\text{DISTRIBUTED}\\ \text{EDGE LOADS}\\ \text{SECTION}\end{array}} \rightarrow \text{ELOAD(NELEM,NEVAB)}]\text{C(LGDATA)}$$

C(LGDATA)
$$\begin{bmatrix} \text{COORD(NPOIN,NDIME)} \\ \text{LNODS(NELEM,NNODE)} \\ \text{WEIGP(NGAUS)} \\ \text{POSGP(NGAUS)} \end{bmatrix}$$

C(CONTRO)
$$\begin{bmatrix} \text{NDOFN} \\ \text{NDIME} \\ \text{NGAUS} \end{bmatrix}$$

C(WORK)
$$\begin{bmatrix} \text{SHAPE(NNODE)} \\ \text{DERIV(NDIME,NNODE)} \end{bmatrix}$$

Annotated FORTRAN listing

```
C
C*** DISTRIBUTED EDGE LOADS SECTION
C
      READ(5,930) NEDGE                                  ⎫
  930 FORMAT(I5)                                         ⎪
      WRITE(6,935) NEDGE                                 ⎬  1*
  935 FORMAT(1H0,5X,21HNO. OF LOADED EDGES =,I5)         ⎪
      WRITE(6,940)                                       ⎭
  940 FORMAT(1H0,5X,
     . 38HLIST OF LOADED EDGES AND APPLIED LOADS)
      NODEG=3                                                2*
C
C*** LOOP OVER EACH LOADED EDGE
C
      DO 160 IEDGE=1,NEDGE                                   3*
C
C*** READ DATA LOCATING THE LOADED EDGE AND
C    APPLIED LOAD
C
      READ(5,945) NEASS,(NOPRS(IODEG),IODEG=            ⎫
     . 1,NODEG)                                         ⎪
  945 FORMAT(4I5)                                       ⎬  4*
      WRITE(6,950) NEASS,(NOPRS(IODEG),IODEG=           ⎪
     . 1,NODEG)                                         ⎭
  950 FORMAT(I10,5X,3I5)
      READ(5,955) ((PRESS(IODEG,IDOFN),IODEG=           ⎫
     . 1,NODEG),IDOFN=1,NDOFN)                          ⎬
      WRITE(6,955) ((PRESS(IODEG,IDOFN),IODEG=          ⎭  5*
```

```
                 , 1,NODEG),IDOFN=1,NDOFN)                         ⎤
         955 FORMAT(6F10.3)                                        ⎦
             ETASP=-1.0                                            6*
     C
     C*** CALCULATE THE COORDINATES OF THE NODES
     C    OF THE ELEMENT EDGE
     C
     ┌────── DO 100 IODEG=1,NODEG                                  ⎫
     │       LNODE=NOPRS(IODEG)                                    ⎬ 7*
     │ ┌──── DO 100 IDIME=1,NDIME                                  ⎭
     └─└ 100 ELCOD(IDIME,IODEG)=COORD(LNODE,IDIME)
     C
     C*** ENTER LOOP FOR LINEAR NUMERICAL INTEGRATION
     C
     ┌────── DO 150 IGAUS=1,NGAUS                                  8*
     │       EXISP=POSGP(IGAUS)                                    9*
     C
     C*** EVALUATE THE SHAPE FUNCTIONS AT THE
     C    SAMPLING POINTS
     C
             CALL SFR2(EXISP,ETASP)                                10*
     C
     C*** CALCULATE COMPONENTS OF THE EQUIVALENT
     C    NODAL LOADS
     C
     ┌────── DO 110 IDOFN=1,NDOFN                                  ⎫
     │       PGASH(IDOFN)=0.0                                      ⎬ 11*
     │       DGASH(IDOFN)=0.0                                      ⎭
     │ ┌──── DO 110 IODEG=1,NODEG
     │ │     PGASH(IDOFN)=PGASH(IDOFN)+PRESS(IODEG,               ⎫
     │ │   , IDOFN)*SHAPE(IODEG)                                   ⎬ 12*
     └─└ 110 DGASH(IDOFN)=DGASH(IDOFN)+ELCOD(IDOFN,               ⎫
             , IODEG)*DERIV(1,IODEG)                               ⎬ 13*
             DVOLU=WEIGP(IGAUS)                                    14*
             PXCOM=DGASH(1)*PGASH(2)-DGASH(2)*PGASH(1)            ⎫
             PYCOM=DGASH(1)*PGASH(1)+DGASH(2)*PGASH(2)            ⎬ 15*
     C
     C*** ASSOCIATE THE EQUIVALENT NODAL EDGE
     C    LOADS WITH AN ELEMENT
     C
     ┌────── DO 120 INODE=1,NNODE                                  ⎫
     │       NLOCA=LNODS(NEASS,INODE)                              ⎬ 16*
     │       IF(NLOCA.EQ.NOPRS(1)) GO TO 130                       ⎭
     └─ 120 CONTINUE
         130 JNODE=INODE+NODEG-1                                   17*
             KOUNT=0
     ┌────── DO 140 KNODE=INODE,JNODE                              ⎫
     │       KOUNT=KOUNT+1                                         │
     │       NGASH=(KNODE-1)*NDOFN+1                               │
     │       MGASH=(KNODE-1)*NDOFN+2                               │
     │       IF(KNODE.GT.NNODE) NGASH=1                            ⎬ 18*
     │       IF(KNODE.GT.NNODE) MGASH=2                            │
     │       ELOAD(NEASS,NGASH)=ELOAD(NEASS,NGASH)+
```

```
     .  SHAPE(KOUNT)*PXCOM*DVOLU
 140 ELOAD(NEASS,MGASH)=ELOAD(NEASS,MGASH)+
     .  SHAPE(KOUNT)*PYCOM*DVOLU
 150 CONTINUE
 160 CONTINUE                                            } 19*
```

1* Read and write the number of element edges which are subjected to distributed loading

2* Calculate the number of nodes along an element edge ($= 3$ in this case)

3* Enter loop over each loaded element edge

4* Read and write the nodal point numbers defining the element edge under consideration. The numbering sequence must be as described earlier in the section. Also read the number of the element to which the edge loading is associated.

5* Read and write the values of the normal load/unit length at each node forming the edge, followed by the tangential load intensities at the same points

6* By setting $\eta = -1$ (i.e. ETASP in the program) the first three shape functions for two dimensional plane applications become identical to the three shape functions employed in the beam analysis in Chapter 4. This fact can be checked by setting $\eta = -1$ in the first three expressions of (5.4) and comparing with (4.4). This avoids the need to employ the shape functions for the beam element in the plane stress/strain program.

7* Store the coordinates of the nodal points of the element edge in the local array ELCOD(IDIME,IODEG)

8* Enter the loop for numerical integration along the element edge

9* Set up the coordinate ξ_n at the Gaussian sampling points

10* Calculate the values of the shape functions and their derivatives corresponding to ξ_n and $\eta = -1$.

11* Enter loops over number of degrees of freedom per node, NDOFN and over the number of nodes per element face, NODEG. Also zero the two local arrays PGASH and DGASH.

12* Store the normal and tangential load intensity values at the Gaussian point in PGASH(IDOFN)

13* Store the values of $\partial x/\partial \xi$ and $\partial y/\partial \xi$ at the Gaussian point in DGASH(IDOFN)

14* Set DVOLU equal to the Gaussian weighting function

15* Calculate the terms in brackets in (7.16)

16* Search through the element nodal connection numbers until the node corresponding to the first node specifying the loaded edge is found

17* Set up the correct location for the nodal forces in the ELOAD array

18* Calculate the equivalent nodal forces according to (7.16) and store in the correct locations in array ELOAD. If the element nodal counter KNODE is greater than NNODE then the nodal connection number concerned, must in fact be the first and NGASH and MGASH are respectively set to 1 and 2. This accounts for the case when a distributed load is applied to the face defined by the 7th, 8th and 1st nodal connection numbers. It should be noted that the array ELOAD contains a running total of the equivalent nodal forces. Thus those calculated due to pressure in this section are added to those arising from other sources such as gravity, etc.

19* Termination of DO LOOPS for numerical integration and total number of element edges respectively.

7.4 Thermal loading of solids under plane stress or plane strain conditions

Stresses can often be induced in solids by the presence of a known temperature field. Such stresses can readily be accommodated by use of (1.32) as thermal effects are only a particular case of initial straining. The initial stresses σ^0 corresponding to the initial thermal strains are firstly calculated and converted to equivalent nodal forces in accordance with (1.31). The initial stresses themselves are also stored for subsequent addition to the stresses arising from other sources (e.g. applied loads, etc.) as required by (1.32).

At this stage, differences occur in the corresponding expressions for plane stress situations and plane strain situations. Both cases will be considered, with treatment being carried out in parallel. The first step is to evaluate the initial strains associated with a temperature rise.

Plane stress
For plane stress situations the initial strains (or eigenstrains) are simply [1]

$$\varepsilon_x^0 = \alpha T$$

$$\varepsilon_y^0 = \alpha T$$

$$\gamma_{xy}^0 = 0, \tag{7.17}$$

where α is the coefficient of linear thermal expansion and T is the temperature measured from an arbitrary datum. It is also required for plane stress situations that the initial stress component in the through-thickness direction, σ_z^0, must be zero.

Plane strain

This condition is slightly complicated by the condition in the through-thickness direction, z. In this situation the through thickness initial stress σ_z^0 is non-zero but the corresponding strain component ε_z^0 is required to vanish. Consequently

$$\varepsilon_x^0 = -\frac{v\sigma_z^0}{E} + \alpha T$$

$$\varepsilon_y^0 = -\frac{v\sigma_z^0}{E} + \alpha T$$

$$\gamma_{xy}^0 = 0$$

$$\varepsilon_z^0 = 0 = \frac{\sigma_z^0}{E} + \alpha T, \tag{7.18}$$

where v is the Poisson's ratio for the material. Using the last expression to eliminate σ_z^0 we have

$$\varepsilon_x^0 = (1 + v)\alpha T$$

$$\varepsilon_y^0 = (1 + v)\alpha T$$

$$\gamma_{xy}^0 = 0, \tag{7.19}$$

and

$$\sigma_z^0 = -E\alpha T. \tag{7.20}$$

The portion of subroutine LOADPS which calculates the initial stresses σ^0 and the nodal forces equivalent to these, as given by (1.31), can now be assembled and will be termed the THERMAL LOADING SECTION.

The values of the four components of $\sigma^0 = [\sigma_x^0, \sigma_y^0, \tau_{xy}^0, \sigma_z^0]^T$ at each Gaussian point will be stored in the array

STRIN(JSTRE,KGAST),

where the index JSTRE ranges from 1 to 4 and KGAST ranges from 1 to NELEM*NGAUS*NGAUS i.e. the total number of Gaussian integration points in the structure.

Input/Output diagram

	INPUT		OUTPUT	
F(5)	$\begin{bmatrix} \text{ITEMP} \\ \text{TEMPE(NPOIN)} \end{bmatrix} \rightarrow$	THERMAL LOADING SECTION	$\rightarrow$ ELOAD(NELEM,NEVAB)] C(LGDATA)	

$$C(CONTRO) \begin{bmatrix} NPOIN \\ NELEM \\ NNODE \\ NDIME \\ NGAUS \end{bmatrix}$$

$$C(LGDATA) \begin{bmatrix} COORD(NPOIN,NDIME) \\ MATNO(NELEM) \\ LNODS(NELEM,NNODE) \\ WEIGP(NGAUS) \\ POSGP(NGAUS) \\ PROPS(NMATS,NPROP) \end{bmatrix}$$

$$C(WORK) \begin{bmatrix} DMATX(NSTRE,NSTRE) \\ SHAPE(NNODE) \\ CARTD(NDIME,NNODE) \end{bmatrix}$$

A(JACOB2) [DJACB

Annotated FORTRAN listing

```
      C
      C*** THERMAL LOADING SECTION
      C
      C
      C*** INITIALIZE AND INPUT THE NODAL TEMPERATURES
      C
            DO 170 IPOIN=1,NPOIN                              }  1*
        170 TEMPE(IPOIN)=0.0
            WRITE(6,960)
        960 FORMAT(1H0,5X,
           . 29HPRESCRIBED NODAL TEMPERATURES)               }  2*
        180 READ(5,965) NODPT,TEMPE(NODPT)
            WRITE(6,965) NODPT,TEMPE(NODPT)
        965 FORMAT(I5,F10.3)
            IF(NODPT.LT.NPOIN) GO TO 180                         3*
            MDIME=NDIME+1                                        4*
            KGAST=0
      C
      C*** LOOP OVER EACH ELEMENT
      C
            DO 280 IELEM=1,NELEM                                 5*
            LPROP=MATNO(IELEM)                                  6*
            DO 200 INODE=1,NNODE
            LNODE=LNODS(IELEM,INODE)
      C
      C*** IDENTIFY THE COORDINATES AND TEMPERATURE
      C     OF EACH ELEMENT NODE POINT
      C
            DO 190 IDIME=1,NDIME                              }  7*
        190 ELCOD(IDIME,INODE)=COORD(LNODE,IDIME)
        200 ELCOD(MDIME,INODE)=TEMPE(LNODE)                     8*
      C
      C*** SET UP MATERIAL PROPERTIES
      C
```

```
      CALL MODPS(LPROP)                                     9*
      YOUNG=PROPS(LPROP,1)                                 10*
      POISS=PROPS(LPROP,2)                                 11*
      THICK=PROPS(LPROP,3)                                 12*
      ALPHA=PROPS(LPROP,5)                                 13*
C
C*** ENTER LOOPS FOR AREA NUMERICAL INTEGRATION
C
      DO 270 IGAUS=1,NGAUS                              }  14*
      DO 270 JGAUS=1,NGAUS
      KGAST=KGAST+1                                        15*
      EXISP=POSGP(IGAUS)                               }  16*
      ETASP=POSGP(JGAUS)
C
C*** EVALUATE THE SHAPE FUNCTIONS AND TEMPERATURE
C     AT THE SAMPLING POINTS,ELEMENTAL VOLUME
C     AND CARTESIAN DERIVATIVES
C
      CALL SFR2(EXISP,ETASP)                              17*
      KGASP=1                                             18*
      CALL JACOB2(IELEM,DJACB,KGASP)                      19*
      THERM=0.0
      DO 210 INODE=1,NNODE                             }  20*
  210 THERM=THERM+ELCOD(MDIME,INODE)*SHAPE(INODE)
      DVOLU=DJACB*WEIGP(IGAUS)*WEIGP(JGAUS)               21*
      IF(THICK.NE.0.0) DVOLU=DVOLU*THICK                  22*
C
C*** EVALUATE THE INITIAL THERMAL STRAINS
C
      EIGEN=THERM*ALPHA                                   23*
      IF(NTYPE.EQ.2) GO TO 220                            24*
      STRAN(1)=-EIGEN
      STRAN(2)=-EIGEN                                  }  25*
      STRAN(3)=0.0
      GO TO 230
  220 STRAN(1)=-(1.0+POISS)*EIGEN
      STRAN(2)=-(1.0+POISS)*EIGEN                      }  26*
      STRAN(3)=0.0
C
C*** AND THE CORRESPONDING INITIAL STRESSES
C
  230 DO 250 ISTRE=1,NSTRE                             }
      STRES(ISTRE)=0.0
      DO 240 JSTRE=1,NSTRE                             }  27*
  240 STRES(ISTRE)=STRES(ISTRE)+DMATX(ISTRE,
     .  JSTRE)*STRAN(JSTRE)
  250 STRIN(ISTRE,KGAST)=STRES(ISTRE)                     28*
      IF(NTYPE.EQ.2) STRIN(4,KGAST)=-YOUNG*EIGEN          29*
      IF(NTYPE.EQ.1) STRIN(4,KGAST)=0.0                   30*
C
C*** CALCULATE THE EQUIVALENT NODAL FORCES AND
C     ASSOCIATE WITH THE ELEMENT NODES
C
```

```
    ┌────DO 260 INODE=1,NNODE
    │    NGASH=(INODE-1)*NDOFN+1
    │    MGASH=(INODE-1)*NDOFN+2
    │    ELOAD(IELEM,NGASH)=ELOAD(IELEM,NGASH)
    │  . =(CARTD(1,INODE)*STRES(1)+CARTD(2,INODE)*
    │  . STRES(3))*DVOLU
    └260 ELOAD(IELEM,MGASH)=ELOAD(IELEM,MGASH)
       .=(CARTD(1,INODE)*STRES(3)+CARTD(2,INODE)*
       . STRES(2))*DVOLU
   270 CONTINUE
   280 CONTINUE
```

$\left.\begin{array}{c} \\ \\ \end{array}\right\}$ 31 *

$\left.\begin{array}{c} \\ \\ \end{array}\right\}$ 32 *

1* Set all nodal temperatures to zero
2* Read and write the non-zero nodal temperatures. The temperature value at the highest numbered node must be input whether it is zero or not
3* Set MDIME = 3. This index will be used to locate the temperature term in subsequent arrays
4* Set KGAST = 0. This index will locate the number of the Gaussian integration point for the later specification of the initial stresses at each such point
5* Enter loop over each element
6* Identify the material property type of the element
7* Store the coordinates of the nodal points of the element under consideration in the local array ELCOD(IDIME,INODE)
8* Store the temperatures at the nodal point of the element in ELCOD (MDIME,INODE)
9* Call subroutine MODPS(LPROP) which calculates the **D** matrix for the element material type as identified by LPROP
10* Set YOUNG equal to the material elastic modulus
11* Set POISS equal to the Poisson's ratio
12* Set THICK equal to the element thickness
13* Set ALPHA equal to the coefficient of thermal expansion
14* Enter loops for Gaussian integration over the element area
15* Increment KGAST to give the value for the current Gauss point
16* Set up the coordinates ξ_n, η_m of the Gaussian sampling points
17* Evaluate the shape functions N_i and the derivatives $\partial N_i/\partial \xi$, $\partial N_i/\partial \eta$ corresponding to ξ_n, η_m
18* Set KGASP = 1 merely to avoid overstepping the bounds of array GPCOD in subroutine JACOB2. Array GPCOD is not used in the present calculations.
19* Calculate the Jacobian matrix **J**, its determinant, DJACB, its inverse $[\mathbf{J}]^{-1}$ and the Cartesian derivatives of the shape functions CARTD (IDIME,INODE). The index IDIME ranges from 1 to 2 and relates

to the x and y derivatives respectively and INODE ranges from 1 to 8 denoting the 8 shape functions of the parabolic isoparametric element.

20* Calculate the temperature value at the Gauss point, and store as THERM.

21* Calculate the equivalent of the elemental volume DVOLU, for numerical integration

22* If the material thickness is non-zero multiply DVOLU by the thickness, i.e. Default value of the element thickness is assumed to be unity

23* Set EIGEN $= \alpha T$

24* Branch for plane stress or plane strain situations

25* Calculate the initial strains STRAN(I) for plane stress problems according to (7.17)

26* Calculate the initial strains STRAN(I) for plane strain problems according to (7.18)

27* Calculate the initial stresses corresponding to the strains according to $\boldsymbol{\sigma}^0 = \mathbf{D}\boldsymbol{\varepsilon}^0$. These are held in the temporary array STRES(ISTRE)

28* Also store the initial stresses in the array STRIN(ISTRE,KGAST) i.e. store the initial stresses for each Gauss point

29* For plane strain problems evaluate σ_z^0 according to (7.20)

30* For plane stress problems set $\sigma_z^0 = 0$

31* Calculate the equivalent nodal forces according to (1.31) and store in the correct locations in the ELOAD array. The matrix multiplication in (1.31) has been condensed as follows

$$
\begin{bmatrix} F_{xi} \\ F_{yi} \end{bmatrix} = -\int_{V_e} [\mathbf{B}_i]^\mathrm{T} \boldsymbol{\sigma}^0 \, \mathrm{d}V = -\int_{V_e} \begin{bmatrix} \dfrac{\partial N_i}{\partial x} & 0 & \dfrac{\partial N_i}{\partial y} \\ 0 & \dfrac{\partial N_i}{\partial y} & \dfrac{\partial N_i}{\partial x} \end{bmatrix} \begin{bmatrix} \sigma_x^0 \\ \sigma_y^0 \\ \tau_{xy}^0 \end{bmatrix} \mathrm{d}V
$$

$$
= -\int_{V_e} \begin{bmatrix} \dfrac{\partial N_i}{\partial x}\sigma_x^0 + \dfrac{\partial N_i}{\partial y}\tau_{xy}^0 \\ \dfrac{\partial N_i}{\partial y}\sigma_y^0 + \dfrac{\partial N_i}{\partial x}\tau_{xy}^0 \end{bmatrix} \mathrm{d}V. \qquad (7.21)
$$

32* Termination of DO LOOPS for numerical integration and total number of elements respectively.

7.5 Nodal point loads and assembly of loading subroutine

The only form of loading which has not yet been considered is that of discrete loading at nodal points. This can easily be accommodated by inputting directly into the ELOAD array. A slight complication arises in that a load may possibly act at a node to which several elements are connected. Since the ELOAD array contains the nodal forces for each element separately, it is necessary to associate the nodal force with any one of the attached elements arbitrarily. It then remains to combine the previous sections of the loading subroutine and to input some control data which specifies what forms of loading are to be considered for any particular problem.

This control information consists of four parameters IPLOD, IGRAV, IEDGE and ITEMP which respectively control the input of point loads, gravity loads, distributed loads on element edges and thermal loading. If a zero value for any of these parameters is input then it is implied that the form of loading relating to the particular parameter is not to be input for this loading case. If a unit value is input then loading is to be expected. For example, if the four values are input as

$$0 \quad 1 \quad 0 \quad 1$$

it means that gravity and thermal loading only is to be considered for this load case. The subroutine can now be completed and assembled.

Input/Output diagram

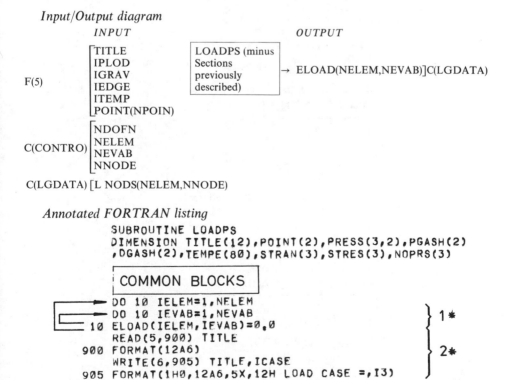

Annotated FORTRAN listing

```
      SUBROUTINE LOADPS
      DIMENSION TITLE(12),POINT(2),PRESS(3,2),PGASH(2)
     ,DGASH(2),TEMPE(80),STRAN(3),STRES(3),NOPRS(3)
```

```
C
C*** READ DATA CONTROLLING LOADING TYPES
C     TO BE INPUT
C
      READ(5,910) IPLOD,IGRAV,IEDGE,ITEMP
      WRITE(6,910) IPLOD,IGRAV,IEDGE,ITEMP          } 3*
  910 FORMAT(4I5)
C
C*** READ NODAL POINT LOADS
C
      IF(IPLOD.EQ.0) GO TO 500                         4*
   20 READ(5,915) LODPT,(POINT(IDOFN),IDOFN=
     . 1,NDOFN)
      WRITE(6,915) LODPT,(POINT(IDOFN),IDOFN=        } 5*
     . 1,NDOFN)
  915 FORMAT(I5,2F10.3)
C
C*** ASSOCIATE THE NODAL POINT LOADS WITH
C     AN ELEMENT
C
      DO 30 IELEM=1,NELEM
      DO 30 INODE=1,NNODE
      NLOCA=LNODS(IELEM,INODE)
      IF(LODPT.EQ.NLOCA) GO TO 40                    } 6*
   30 CONTINUE
   40 DO 50 IDOFN=1,NDOFN
      NGASH=(INODE-1)*NDOFN+IDOFN
   50 ELOAD(IELEM,NGASH)=POINT(IDOFN)
      IF(LODPT.LT.NPOIN) GO TO 20                      7*
  500 CONTINUE
      IF(IGRAV.EQ.0) GO TO 600                         8*
```

GRAVITY LOADING   9*

```
  600 CONTINUE
      IF(IEDGE.EQ.0) GO TO 700                        10*
```

DISTRIBUTED EDGE LOADING   11*

```
  700 CONTINUE
      IF(ITEMP.EQ.0) GO TO 800                        12*
```

TEMPERATURE LOADING   13*

```
  800 CONTINUE
      WRITE(6,970)
  970 FORMAT(1H0,5X,
```

```
          . 36H TOTAL NODAL FORCES FOR EACH ELEMENT)
     ┌────── DO 290 IELEM=1,NELEM                                      ⎫
     └─── 290 WRITE(6,975) IELEM,(ELOAD(IELEM,IEVAB),              ⎬ 14*
          . IEVAB=1,NEVAB)                                             ⎭
      975 FORMAT(1X,I4,5X,8E12.4/(10X,8E12.4))
          RETURN
          END
```

1* Set to zero the array ELOAD in which the equivalent nodal forces for each element are to be stored

2* Read and write the load case title (limited to 72 alphanumeric characters). Also write the load case number as defined by ICASE, the loop parameter over the total number of load cases to be solved for. This index is set in the master subroutine to be developed in Chapter 10.

3* Read and write the loading control parameters described previously

4* If no point loads are to be input, go to 500

5* Read and write the node number and the x and y components of load for any nodal point at which discrete loads are applied

6* Associate the nodal point loads with any one element connected to the relevant nodal point

7* The last nodal point number must be input whether it is loaded or not

8* If no gravity effects are to be considered, go to 600

9* Insert the GRAVITY SECTION listed in Section 7.2

10* If there are no element edges where distributed loads are applied, go to 700

11* Insert the DISTRIBUTED EDGE LOADS SECTION listed in Section 7.3

12* If thermal loading is not to be considered go to 800

13* Insert the THERMAL LOADING SECTION listed in Section 7.4

14* Write the total nodal forces for each element

On exit from subroutine LOADPS the nodal forces are ready for subsequent insertion into the solution subroutine to be described in Chapter 8.

7.6 Loading data for plate bending situations

For plate bending applications two forms of loading will be considered. Firstly load components corresponding to the permissible generalised forces may be prescribed at the nodal points. Thus with respect to Fig. 7.4, at each node a load in the z direction and couples acting in both the xz and yz planes may be input. Secondly a uniformly distributed load acting normal to the plate (i.e. in the z direction) may be applied. As in the previous sections such a loading must be converted into equivalent nodal forces before

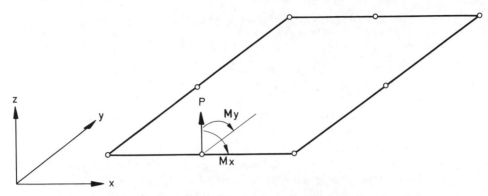

FIG. 7.4.   Generalised nodal forces for a plate bending element.

equation solution takes place. This can be readily achieved by use of the general expression (1.29) which for the present situation becomes

$$\mathbf{F}_i = \begin{bmatrix} P \\ M_x \\ M_y \end{bmatrix}_i = \int_{A_e} N_i \begin{bmatrix} q \\ 0 \\ 0 \end{bmatrix} \mathrm{d}A, \qquad (7.22)$$

where q is the distributed load intensity and integration is taken over the element area. The subroutine which evaluates these equivalent nodal forces and input data for discrete nodal loading can now be formulated.

Input/Output diagram

	INPUT	OUTPUT	
Γ(5)	⎡TITLE IPLOD ⎣POINT(NPOIN)	→ LOADPB → FLOAD(NELEM,NEVAB)	⎤C(LGDATA) ⎦

C(CONTRO) ⎡NDOFN
NELEM
NEVAB
⎣NNODE

C(LGDATA) ⎡LNODS(NELEM,NNODE,COORD(NPOIN,NDIME),MATNO(NELEM),
⎣WEIGP(NGAUS),POSGP(NGAUS),PROPS(NMATS,NPROP)

Annotated FORTRAN listing

```
        SUBROUTINE LOADPB
C
C*** CALCULATE NODAL FORCES FOR PLATE ELEMENT
C
        DIMENSION TITLE(12),POINT(3)
```

COMMON BLOCKS

```
         DO 10 IELEM=1,NELEM
         DO 10 IEVAB=1,NEVAB                                   } 1*
         ELOAD(IELEM,IEVAB)=0.0
      10 CONTINUE
         READ(5,900) TITLE
     900 FORMAT(12A6)                                          } 2*
         WRITE(6,905) TITLE,ICASE
     905 FORMAT(1H0,12A6,5X,12H LOAD CASE =,I3)
C
C*** READ DATA CONTROLLING LOADING
C    TYPES TO BE INPUT
C
         READ(5,910) IPLOD
     910 FORMAT(I5)                                            } 3*
         WRITE(6,915) IPLOD
     915 FORMAT(I5)
C
C*** READ NODAL POINT LOADS
C
         IF(IPLOD.EQ.0) GO TO 60                               4*
      20 READ(5,920)
        .LODPT,(POINT(IDOFN),IDOFN=1,NDOFN)
         WRITE(6,925)
        .LODPT,(POINT(IDOFN),IDOFN=1,NDOFN)                    } 5*
     920 FORMAT(I5,3F10.5)
     925 FORMAT(I5,3F10.5)
C
C*** ASSOCIATE THE NODAL POINT LOADS
C    WITH AN ELEMENT
C
         DO 30 IELEM=1,NELEM
         DO 30 INODE=1,NNODE
         NLOCA=LNODS(IELEM,INODE)
         IF(LODPT.EQ.NLOCA) GO TO 40
      30 CONTINUE
      40 DO 50 IDOFN=1,NDOFN                                   } 6*
         NGASH=(INODE-1)*NDOFN+IDOFN
         ELOAD(IELEM,NGASH)=POINT(IDOFN)
      50 CONTINUE
         IF(LODPT.NE.NPOIN) GO TO 20
      60 CONTINUE
C
C*** LOOP OVER EACH ELEMENT
C
         DO 110 IELEM=1,NELEM                                  7*
         LPROP=MATNO(IELEM)                                    8*
         UDLOD=PROPS(LPROP,4)                                  9*
         IF(UDLOD.EQ.0.0) GO TO 110                            10*
C
C*** EVALUATE THE COORDINATES OF THE
C    ELEMENT NODAL POINTS
C
```

```
        ┌──► DO 70 INODE=1,NNODE
        │    LNODE=LNODS(IELEM,INODE)           ⎫
     ┌──┼──► DO 70 IDIME=1,NDIME                ⎬ 11*
     │  │    ELCOD(IDIME,INODE)=COORD(LNODE,IDIME) ⎭
     └─ 70 CONTINUE
        ┌──► DO 80 IEVAB=1,NEVAB                ⎫
        │    ELOAD(IELEM,IEVAB)=0.0             ⎬ 12*
        └─ 80 CONTINUE                          ⎭
             KGASP=0                            13*
C
C*** ENTER LOOPS FOR NUMERICAL INTEGRATION
C
     ┌────► DO 100 IGAUS=1,NGAUS
     │      EXISP=POSGP(IGAUS)                  ⎫
     │  ┌─► DO 100 JGAUS=1,NGAUS                ⎬ 14*
     │  │   ETASP=POSGP(JGAUS)                  ⎭
     │  │   KGASP=KGASP+1                       15*
C
C*** EVALUATE THE SHAPE FUNCTIONS AT THE
C    SAMPLING POINTS AND ELEMENTAL AREA
C
            CALL SFR2(EXISP,ETASP)              16 *
            CALL JACOB2(IELEM,DJACB,KGASP)      17 *
            DAREA=DJACB*WEIGP(IGAUS)*WEIGP(JGAUS) 18 *
C
C*** CALCULATE LOADS AND ASSOCIATE WITH
C    ELEMENT NODAL POINTS
C
     │  │┌─► DO 90 INODE=1,NNODE                ⎫
     │  ││  NPOSN=(INODE-1)*NDOFN+1             ⎬ 19*
     │  ││  ELOAD(IELEM,NPOSN)=ELOAD(IELEM,NPOSN)+
     │  ││  .SHAPE(INODE)*UDLOD*DAREA           ⎭
     │  │└─ 90 CONTINUE
     │  └─ 100 CONTINUE                         ⎫ 20*
     └──── 110 CONTINUE                         ⎭
             WRITE(6,930)
        930 FORMAT(1H0,5X,
             .36H TOTAL NODAL FORCES FOR EACH ELEMENT)
     ┌────► DO 120 IELEM=1,NELEM                ⎫ 21*
     │      WRITE(6,935) IELEM,                 ⎬
     │      .(ELOAD(IELEM,IEVAB),IEVAB=1,NEVAB) ⎭
     └──── 120 CONTINUE
        935 FORMAT(1X,I4,5X,8E12.4/(10X,8E12.4)/
             .(10X,8E12.4))
             RETURN
             END
```

1* Set to zero the array ELOAD in which the equivalent nodal forces are to be stored. It should be remembered that for plate bending problems the number of degrees of freedom per node, NDOFN = 3

2* Read and write the load case title. Also write the load case number as defined by ICASE in the master subroutine

3* Read and write the control parameter IPLOD. If IPLOD is specified as zero no discrete nodal loads or couples are to be input

4* If no nodal loads or couples are to be input, go to 60

5* Read and write the node number, the nodal load normal to the plate and the couples acting in the xz and yz planes respectively (see Fig. 7.4). Data for the last nodal point number must be input, whether it is loaded or not.

6* Associate the nodal point loads with an element and store in the element load array ELOAD

7* Enter loop over each element

8* Identify property set number for the current element

9* Set UDLOD equal to the uniformly distributed load intensity

10* If this is zero, proceed to the next element

11* Store the coordinates of the nodal points of the element in the local array ELCOD(IDIME,INODE)

12* Set ELOAD to zero

13* Set Gauss point counter to zero

14* Enter the loops for Gaussian integration over the element area and set up the coordinates (ξ_n, η_m) of the Gaussian sampling points for use in (7.22)

15* Increment Gauss point counter by one

16* Evaluate the shape functions N_i and the derivatives $\partial N_i/\partial \xi$, $\partial N_i/\partial \eta$ for coordinate values ξ_n, η_m

17* Calculate the Jacobian matrix $\mathbf{J}$, its determinant, DJACB, its inverse $[\mathbf{J}]^{-1}$ and the Cartesian derivatives of the shape functions

18* Calculate the elemental area $dA \times$ (the integrating constants)

19* Calculate the equivalent nodal forces according to (7.22) and store in the ELOAD array

20* Termination of DO LOOPS for numerical integration, total number of elements and total number of different materials respectively

21* Write the total generalised nodal forces for each nodal point

On exit from subroutine LOADPB the generalised nodal forces have been evaluated and stored ready for utilisation in the solution routine.

References

1. Timoshenko, S., and Goodier, J. N., "Theory of Elasticity". McGraw-Hill, New York, 1951.

8

The Equation Solution Subroutine

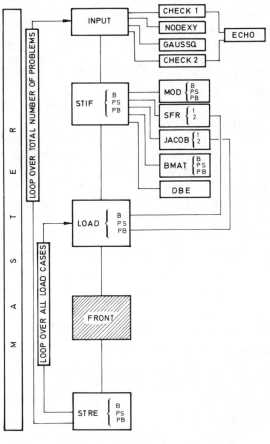

8.1 Introduction

The method adopted for equation solution is a major factor influencing the efficiency of any finite element program. Several options are open to the programmer ranging from iterative methods such as the Gauss–Seidel technique [1] to the direct Gaussian elimination algorithms. In this text we shall employ a direct elimination process and in particular the frontal method of equation assembly and reduction will be described. The frontal equation solution technique was originated by Irons [2] and has earned the reputation of being easy and inexpensive to use. A direct Gaussian elimination algorithm was described in Chapter 2 and the reader may find it useful to review that chapter before proceeding with this chapter.

The frontal method can be considered as a particular technique for first assembling finite element stiffnesses and nodal forces into a global stiffness matrix and load vector and then solving for the unknown displacements by means of a Gaussian elimination and backsubstitution process. It is designed to minimise core storage requirements, the number of arithmetic operations and the use of peripheral equipment. Its efficiency and advantages over other algorithms, such as banded equation techniques, will become apparent during the description of the method and will be summarised later in Section 8.8 of this chapter. In the original frontal program [2], equation solution and other operations, such as element stiffness formation, were performed in a single integrated program. In this chapter we will describe a frontal solver in subroutine form which allows the solution process to be separated entirely from other subroutines. It is to be hoped that this modular approach facilitates the understanding of the method and the development of the programs. The format of the chapter is to describe the frontal solution method, to give the program listing and finally, to summarise the technique.

8.2 General description of the frontal technique

The frontal technique as described in this chapter is applicable only to the solution of symmetric systems of linear stiffness equations. The main idea of the frontal solution is to *assemble the equations and eliminate the variables at the same time*. As soon as the coefficients of an equation are completely assembled from the contributions of all relevant elements, the corresponding variable can be eliminated. Therefore the complete structural stiffness matrix is never formed as such, since after elimination the reduced equation is immediately transferred to back-up disc storage.

The core contains, at any given instant, the upper triangular part of a square matrix containing the equations which are being formed at that particular time. These equations, their corresponding nodes and degrees of freedom are termed the *front*. The number of unknowns in the front is the *frontwidth*; this length generally changes continually during the assembly/reduction process. The maximum size of problem which can be solved is governed by the *maximum frontwidth*. The equations, nodes and degrees of freedom belonging to the front are termed *active*; those which are yet to be considered are *inactive*; those which have passed through the front and have been eliminated are said to be *deactivated*.

During the assembly/elimination process the elements are considered each in turn according to a prescribed order. Whenever a new element is called in, its stiffness coefficients are read from disc file and summed either into existing equations, if the nodes are already active, or into new equations which have to be included in the front if the nodes are being activated for the first time. If some nodes are appearing for the last time, the corresponding equations can be eliminated and stored away on a disc file and are thus deactivated. In so doing they free space in the front which can be employed during assembly of the next element.

More details concerning the technique are presented in Sections 8.3 to 8.9. Readers who wish to merely treat the solution routine as a "black box" can proceed immediately to Chapter 9.

8.3 The "Pre-front" section of the solution routine

8.3.1 Prescribed displacement data

Since each equation to be assembled and eliminated corresponds to a particular degree of freedom rather than a node number it is necessary to rearrange that data which relates to prescribed displacements. The fixity code and prescribed displacement values are transferred to vector arrays extending over the total number of degrees of freedom in the structure. The fixity integers are transferred from the IFPRE array, described in Chapter 3, to the array IFFIX(IPOSN). Index IPOSN determines the position of a particular nodal degree of freedom in the vector. The IDOFNth degree of freedom of node IPOIN is located according to IPOSN $= (IPOIN-1)*NDOFN + IDOFN$. The array is first zeroed and the unit code values are then inserted in the correct locations.

Similarly the prescribed displacement values are transferred from the PRESC array to array FIXED(IPOSN), where the location of the prescribed values is identical to the previous case.

8.3.2 Registration of the last appearance of each node

In order to know when a nodal variable can be eliminated it is essential that the last appearance of a node in the element nodal point listings be recorded in some way. This is achieved by means of a loop over each element nested within an outer loop over each nodal point. For a particular nodal point number, IPOIN, a search is conducted over the elements in the order in which they are to be processed in the element topology listing, LNODS(IELEM,INODE), its location is recorded by storing the element number IELEM and the nodal position INODE as KLAST and NLAST respectively. This is done for each element, with successive identifications of the node in question being stored in the original variables KLAST and NLAST by overwriting. Hence on completion of the search through all the elements, the final appearance of the node is located as LNODS(KLAST, NLAST). To register this final appearance a negative sign is placed in the LNODS entry before the corresponding nodal identification number, IPOIN. This is repeated for each nodal point number in turn.

8.4 Formulation of the destination, active variables and eliminated variables arrays

As previously mentioned the frontal process operates by performing assembly and elimination as joint operations. By eliminating variables as soon as their assembly has been completed, core storage is made available for nodal variables yet to be assembled. To enable this process to function a relatively elaborate "housekeeping" system is required. In particular the position in the front into which each degree of freedom of an element is to be assembled must be recorded as well as a list of the active variables currently in the front.

8.4.1. The destination vector

The assembly position in the front of each degree of freedom of a given element is monitored by use of the array NDEST(KDOFE), where KDOFE ranges over the total number of degrees of freedom per element. This indicates that the KDOFEth degree of freedom (as defined by the nodal connection numbers) of the element under consideration is to be located in the NDEST(KDOFE) position in the front. Since it defines the destination of the variable it is known as the *destination vector*. The destination vector is a local array and is computed separately on the introduction of each element.

8.4.2 The variable location vector

Each element nodal variable is first assigned a location according to the following algorithm

$$LOCEL(NPOSI) = (LNODS(IELEM,INODE) - 1)*NDOFN + IDOFN$$

On the right hand side of the expression we state that the IDOFNth degree of freedom of the INODEth node of a particular element IELEM is assigned a location increasing with nodal point numbering and with increasing number of degrees of freedom. The left hand side indicates that this information is stored in the vector LOCEL; position within the array being defined by

$$NPOSI = (INODE - 1)*NDOFN + IDOFN,$$

where INODE defines the element nodal point in question and IDOFN relates to the nodal degree of freedom. It should be noted that location within LOCEL is done on an element level whereas the degrees of freedom of the element are assigned identification positions on a global basis, i.e. an entry in LOCEL may have any number between 1 and NTOTV, where NTOTV is the total number of variables in the structure. Array LOCEL is constructed for each element in turn before the assembly-elimination process begins. If, when considering a particular element, it is detected that the variable is making a final appearance (as registered by a negative sign in the LNODS array) then a negative sign is placed before the corresponding entry in the LOCEL vector also.

8.4.3 The vector of active variables

This vector contains a list of all the variables currently in the front and is named array NACVA(IFRON), where IFRON ranges over the maximum permissible length of the front, MFRON. This states that the IFRONth equation of the front relates to variable NACVA(IFRON) where the variable location is as defined by LOCEL in the previous section. A zero value in NACVA indicates an available space which can be employed in the assembly of later elements. The content of the frontwidth and consequently the NACVA vector is constantly changing as the front is moving through the structure.

The way the NDEST, LOCEL and NACVA vectors are actually employed in solution is best described by means of an illustrative example. Fig. 8.1 shows a hypothetical problem containing 5 nodal points and three triangular elements. Is is assumed that 3 degrees of freedom exist at each nodal point and that the maximum frontwidth, MFRON, allowed in the program is 15. The nodal connection numbers for each element are as specified in Fig. 8.1. The frontal assembly/elimination technique proceeds as follows.

Figure 8.1—Initialisation
The LOCEL, NDEST and NACVA vectors are initially set to zero. At this stage the current frontwidth NFRON is also made zero.

Figure 8.2—Assembly of element 1
Element 1 is introduced for assembly and nodes 4, 1 and 2 become active. The location of the degrees of freedom associated with element 1 are computed and stored in array LOCEL. The vector NACVA is searched for available space for each degree of freedom in turn. Since nothing has been stored yet the 9 degrees of freedom associated with element 1 are stored in the first 9 positions. In particular the degrees of freedom associated with the first nodal connection number (node 4) will be stored in the first three positions in NACVA. These are identified by their global positions 10, 11 and 12, as given by LOCEL. The degrees of freedom associated with node number 1 are preceded by a negative sign in LOCEL to signify that this is the final appearance of node 1 and that it can be eliminated. The current frontwidth NFRON is 9. The destinations in NACVA of the degrees of freedom of element 1 are recorded in the vector NDEST, i.e. degree of freedom 10 occupies the 1st position, degree of freedom 11 occupies the 2nd position and so on.

Figure 8.3—Elimination of degrees of freedom associated with element 1
In Fig. 8.2, vector LOCEL contained negative signs for the degrees of freedom corresponding to node 1. The equations corresponding to these degrees of freedom can now be reduced and the space occupied by them made available for future variables. Thus in NACVA positions 4, 5 and 6 corresponding to the degrees of freedom of node 1 (i.e. 1, 2 and 3) are made zero. Variables 1, 2 and 3 are deactivated. For backsubstitution purposes it must be recorded, on disc file, that the first variables eliminated come from equations 4, 5 and 6 of the front and belong to variable numbers 1, 2 and 3 respectively. As each variable equation is eliminated it is immediately transferred to back-up store. The additional information IFRON, the position of the equation in the front and NIKNO, the number of the variable which has just been eliminated is also transferred to the same disc file. The quantity NIKNO is obtained from vector LOCEL, being detected by the negative signs of components in this array. The frontwidth is unchanged since elimination has taken place of variables in the middle of the front and not at its end.

Figure 8.4—Assembly of element 2
Element 2 is introduced for assembly and nodes 3, 2 and 5 are checked against the list of active nodes 4, 1 and 2. This check reveals that node 2 is already active and that the corresponding equations for its degrees of freedom will be

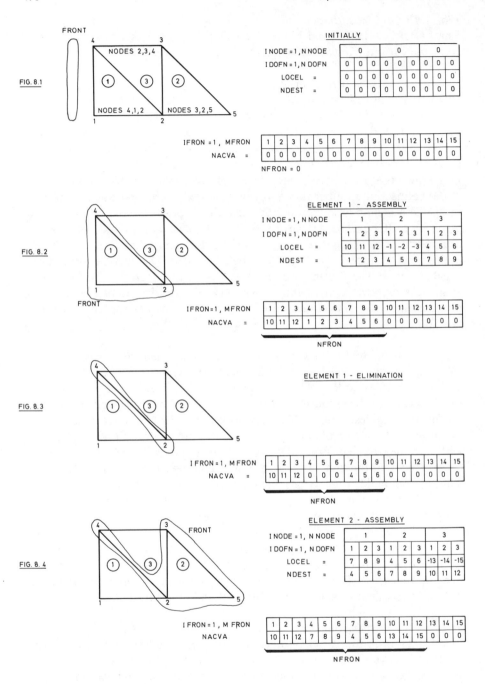

FIG. 8.1

FIG. 8.2

FIG. 8.3

FIG. 8.4

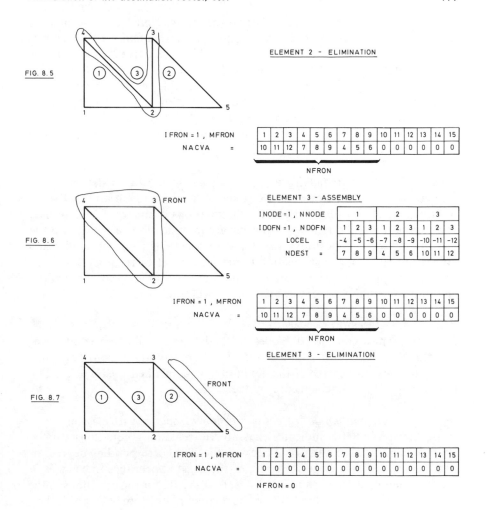

FIG. 8.5

ELEMENT 2 - ELIMINATION

IFRON = 1 , MFRON
NACVA =

1	2	3	4	5	6	7	8	9	10	11	12	13	14	15
10	11	12	7	8	9	4	5	6	0	0	0	0	0	0

NFRON

FIG. 8.6

ELEMENT 3 - ASSEMBLY

INODE =1 , NNODE	1			2			3		
IDOFN =1 , NDOFN	1	2	3	1	2	3	1	2	3
LOCEL =	-4	-5	-6	-7	-8	-9	-10	-11	-12
NDEST =	7	8	9	4	5	6	10	11	12

IFRON = 1 , MFRON
NACVA =

1	2	3	4	5	6	7	8	9	10	11	12	13	14	15
10	11	12	7	8	9	4	5	6	0	0	0	0	0	0

NFRON

ELEMENT 3 - ELIMINATION

FIG. 8.7

IFRON = 1 , MFRON
NACVA =

1	2	3	4	5	6	7	8	9	10	11	12	13	14	15
0	0	0	0	0	0	0	0	0	0	0	0	0	0	0

NFRON = 0

Figs. 8.1–8.7. "Housekeeping" operations for the assembly/elimination process.

the 7th, 8th and 9th in the front. The contributions from node 2, element 2 are then added directly into the existing part of these equations. The degrees of freedom corresponding to nodes 3 and 5 which are active for the first time will be assigned to positions 4, 5, 6 and 10, 11, 12 respectively of the front, where zeroes in NACVA indicated the first available spaces. These destinations are recorded in NDEST. The front will then contain the upper triangular part of a 12 × 12 submatrix of the global stiffness corresponding to the degrees of freedom of the 4 nodal points instantaneously present. The current frontwidth, NFRON, now rises to 12.

Figure 8.5—Elimination of degrees of freedom associated with element 2
The array LOCEL indicates that degrees of freedom 13, 14 and 15 can be deactivated; i.e. the 10th, 11th and 12th equations of the front, will be then complete and can be reduced. Appropriate modifications are made to the NACVA array, with positions 10, 11 and 12 becoming zero. Again the variables eliminated and their position in the front must be recorded for backsubstitution purposes. Since elimination of three variables takes place at the end of the front, the current frontwidth NFRON reduces to 9.

Figure 8.6—Assembly of element 3
Element 3 is now considered. Nodes 2, 3 and 4 are already active and therefore the NACVA vector is not modified. All contributions from element 3 will add to existing parts of the equations as controlled by NDEST. The frontwidth remains unchanged.

Figure 8.7—Elimination of degrees of freedom associated with element 3
Inspection of LOCEL indicates that all the remaining equations can be eliminated and stored away on a back-up file. The corresponding degrees of freedom are deactivated in the order of their last appearance in element 3, i.e. the order of elimination is 7, 8, 9, 4, 5, 6, 10, 11, 12 for the equations in the front and 4, 5, 6, 7, 8, 9, 10, 11, 12 for the degrees of freedom with which they are associated. This information which is recorded in NDEST and LOCEL respectively is also retained for backsubstitution.

   This completes the description of the housekeeping system necessary for the front process. We can now deal with the actual assembly/elimination phase in detail.

8.5 The actual assembly and elimination process

8.5.1 Assembly

For each individual element assembly proceeds in much the same way as in the direct stiffness approach.

The rows of the global stiffness matrix GSTIF and load vector GLOAD are initialised to zero.

The element stiffness matrix, ESTIF, and loads associated with the element, ELOAD, are introduced from disc file and core storage respectively (Generation of the ESTIF array has been discussed in Chapters 4, 5 and 6 and the ELOAD array in Chapter 7).

A pair of nested DO LOOPS is introduced over the total number of degrees of freedom, NEVAB, of the element. The outer loop with index IEVAB ranges from 1 to NEVAB and the inner loop with index JEVAB ranges from 1 to IEVAB. In this way only one half of the element stiffness matrix ESTIF(IEVAB,JEVAB) is ever considered and we become restricted to symmetric matrices immediately.

By means of the destination vector NDEST, we find IDEST and JDEST which represent the global row I and global column J corresponding to the local degrees of freedom IEVAB and JEVAB.

The coefficients ESTIF(IEVAB,JEVAB) are added to GSTIF(IDEST, JDEST)

The loads ELOAD(IEVAB) are added to GLOAD(IDEST).

The only program peculiarity is that, because of symmetry, we need consider only one half of ESTIF and GSTIF and therefore to improve the efficiency of the program the GSTIF array is considered as a one-dimensional array. Figure 8.8 illustrates the storage of a two-dimensional symmetric matrix as a one-dimensional array by considering only the upper or

$$
\begin{bmatrix}
I,J = 1,1 & I,J = 1,2 & I,J = 1,3 & I,J = 1,4 \\
I,J = 2,1 & I,J = 2,2 & I,J = 2,3 & I,J = 2,4 \\
I,J = 3,1 & I,J = 3,2 & I,J = 3,3 & I,J = 3,4 \\
I,J = 4,1 & I,J = 4,2 & I,J = 4,3 & I,J = 4,4
\end{bmatrix}
=
\begin{bmatrix}
N = 1 & N = 2 & N = 4 & N = 7 \\
N = 2 & N = 3 & N = 5 & N = 8 \\
N = 4 & N = 5 & N = 6 & N = 9 \\
N = 7 & N = 8 & N = 9 & N = 10
\end{bmatrix}
$$

FIG. 8.8.   Representation of a two-dimensional symmetric array in vector form.

lower triangle. Conversion from the matrix form $A(I, J)$ to a vector form $A(N)$ is achieved by use of the function

$$
N = \text{NFUNC(I,J)} = \frac{I(I-1)}{2} + J \qquad \text{if } I \geqslant J \text{ (store lower triangle)}
$$

$$N = NFUNC(I,J) = \frac{J(J-1)}{2} + I \qquad \text{if } I \leqslant J \text{ (store upper triangle)}$$

Both functions give the same final vector starting from a symmetric matrix.

8.5.2 Elimination

In the ordinary Gaussian process variables are eliminated in the order in which they are encountered going down the matrix. In the frontal technique the order of elimination is different from the order of formation of the equations. Additionally the order of formation is not straightforward but is governed by the available space in the front. This means that quite often one finds oneself in the position illustrated by the 4×4 example of Fig. 8.9,

$$
\begin{bmatrix}
K_{11} & K_{12} & K_{13} & K_{14} \\
K_{21} & K_{22} & K_{23} & K_{24} \\
K_{31} & K_{32} & K_{33} & K_{34} \\
K_{41} & K_{42} & K_{43} & K_{44}
\end{bmatrix}
\begin{bmatrix}
\delta_1 \\ \delta_2 \\ \delta_3 \\ \delta_4
\end{bmatrix}
=
\begin{bmatrix}
F_1 \\ F_2 \\ F_3 \\ F_4
\end{bmatrix}
$$

FIG. 8.9. Equations for a 4 variable example.

where one has to eliminate first the variable δ_3. The result after elimination is indicated in Fig. 8.10. Apart from row and column 3 the symmetry is maintained and one can keep working on the upper triangle of the front.

$$
\begin{bmatrix}
K_{11}' = K_{11} - \frac{K_{13}}{K_{33}} \cdot K_{31}, & K_{12}' = K_{12} - \frac{K_{13}}{K_{33}} \cdot K_{32}, & 0, & K_{14}' = K_{14} - \frac{K_{13}}{K_{33}} K_{34} \\
K_{21}' = K_{21} - \frac{K_{23}}{K_{33}} \cdot K_{31}, & K_{22}' = K_{22} - \frac{K_{23}}{K_{33}} K_{32}, & 0, & K_{24}' = K_{24} - \frac{K_{23}}{K_{33}} K_{34} \\
K_{31} & K_{32} & , K_{33}, & K_{34} \\
K_{41}' = K_{41} - \frac{K_{43}}{K_{33}} K_{31}, & K_{42}' = K_{42} - \frac{K_{43}}{K_{33}} \cdot K_{32}, & 0, & K_{44}' = K_{44} - \frac{K_{43}}{K_{33}} K_{34}
\end{bmatrix}
\begin{bmatrix}
\delta_1 \\ \delta_2 \\ \delta_3 \\ \delta_4
\end{bmatrix}
=
\begin{bmatrix}
F_1' = F_1 - \frac{K_{13}}{K_{33}} F_3 \\
F_2' = F_2 - \frac{K_{23}}{K_{33}} \cdot F_3 \\
F_3 \\
F_4' = F_4 - \frac{K_{43}}{K_{33}} F_3
\end{bmatrix}
$$

FIG. 8.10.   Reduced equations of Fig. 8.9 after elimination of displacement, δ_3.

However, for backsubstitution purposes the whole of equation 3 is needed. In order to accomplish this and at the same time still only to consider the upper triangle, the terms K_{31} and K_{32} to the left of the diagonal are stored in the "symmetric" positions above the diagonal in locations K_{13} and K_{23} respectively. The whole of row 3 is then transferred to back-up store. Immediately afterwards row 3 is reset to zero to prepare for summation of a new equation which is going to come and occupy the available space. In the program column 3 is also set to zero as would be found on computation anyway.

8.5.3 The case of prescribed displacements

When the elimination reaches a prescribed variable, the process becomes trivial. The right hand sides of the system of equations are modified and the corresponding column of the front matrix (except the diagonal term) is set to zero. If, for example, δ_3 is fixed in the 4×4 example of Fig. 8.9, the system reduces to that shown in Fig. 8.11.

$$
\begin{bmatrix}
K_{11} & K_{12} & 0 & K_{14} \\
K_{21} & K_{22} & 0 & K_{24} \\
K_{31} & K_{32} & K_{33} & K_{34} \\
K_{41} & K_{42} & 0 & K_{44}
\end{bmatrix}
\begin{bmatrix}
\delta_1 \\
\delta_2 \\
\delta_3 \\
\delta_4
\end{bmatrix}
=
\begin{bmatrix}
F_1' = F_1 - K_{13}\delta_3 \\
F_2' = F_2 - K_{23}\delta_3 \\
F_3 + R_3 \\
F_4' = F_4 - K_{43}\delta_3
\end{bmatrix}
$$

Fig. 8.11.   Reduced equations of Fig. 8.9 when displacement δ_3 is prescribed.

Once again, for example, K_{32}, needed for backsubstitution has to be stored in the location of K_{23} since only the upper triangular part of the matrix is retained in the front. Again the whole of row 3 has to be transferred to back-up store.

The nodal force F_3 is the sum of the contributions coming from all elements to which node 3 belongs. *It is not the total nodal force*; in addition to F_3 there must be a concentrated force R_3 applied at this node to produce the prescribed displacements, the value of which can only be found after the backsubstitution phase.

8.5.4 Numerical example

As a numerical example of the techniques described above the same geometry and element topology shown in Fig. 8.1 are used. However, in order to keep the size of the problem within reasonable bounds it is assumed that only one degree of freedom is associated with each nodal point. The evaluation of the effective frontwidth, the frontal global matrix and the frontal load vector during the assembly reduction stages are represented in Fig. 8.12 in analytical

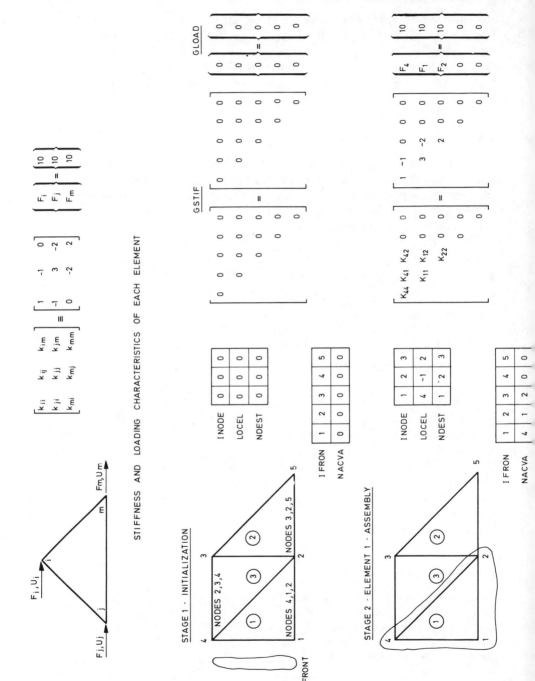

STIFFNESS AND LOADING CHARACTERISTICS OF EACH ELEMENT

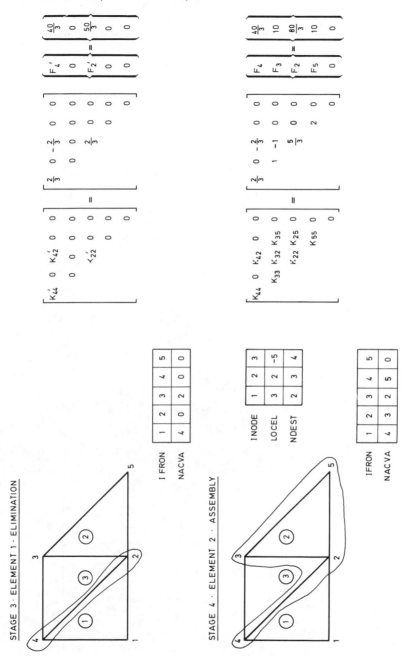

Fig. 8.12(a). The assembly/elimination process.

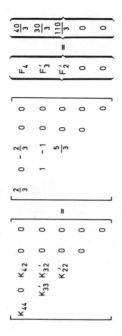

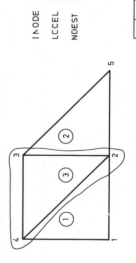

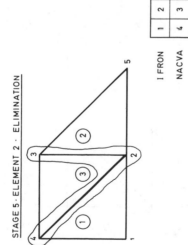

STAGE 5 - ELEMENT 2 - ELIMINATION

STAGE 6 - ELEMENT 3 - ASSEMBLY

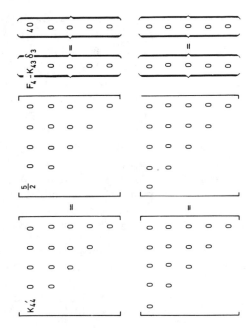

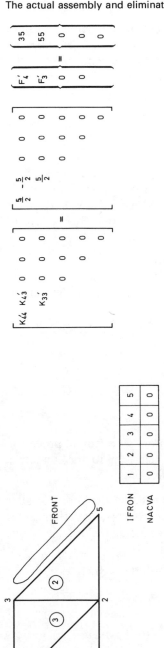

STAGE 7 · ELEMENT 3 · ELIMINATION

STAGE 7 (CONTINUED)

STAGE 7 (CONTINUED)

FIG. 8.12(b). The assembly/elimination process.

and numerical form. The primes in the analytical section indicate values which have just been modified by the elimination process. The numerical values have been obtained assuming that all three elements possess the stiffness and loading characteristics also included in Fig. 8.12. Furthermore it has been assumed that node 3 undergoes a prescribed displacement, $\delta_3 = 2$. The maximum permissible frontwidth, MFRON, is assumed to be 5.

At the end of the assembly/reduction process, a disc file has been created with a total number of records equal to the total number of degrees of freedom in the structure. One record corresponds to an eliminated variable equation, the order of equations being as governed by the front which is in turn governed by the order of element numbering. A record is kept of the order of elimination by means of the variables IFRON and NIKNO. Thus for each eliminated equation we require the information shown in Fig. 8.13 to be transferred to disc file. This is made up as follows:

EQUAT(MFRON)—The reduced frontal equations
EQRHS   —The reduced right hand side loading terms
IFRON   —The position of the eliminated variable in the front at
                            the time of elimination.
NIKNO   —The variable identification name.

This information constitutes one record.

8.6 The backsubstitution phase

8.6.1 Backsubstitution and output of results

The backsubstitution process can best be understood as a frontal process in reverse. The procedure is explained by again referring to the numerical example of Fig. 8.12 considered in the previous section.

Element 3 is reprocessed. The disc file has been left in its final position after assembly/elimination. Information relating to the last equation to be eliminated can then be obtained by simply backspacing the disc file by one

$$\boxed{(\text{EQUAT(LFRON)}, \text{LFRON} = 1, \text{MFRON}), \text{EQRHS}, \text{IFRON}, \text{NIKNO}}$$

FIG. 8.13. Information required for each disc file record.

record. The information displayed in Fig. 8.13 is then available. In particular the eliminated variable NIKNO is known and its position in the front at the time of elimination, IFRON, is also known. The last record will indicate that variable (node) 4 was the last eliminated and occupied the first position in the front. Therefore, one keeps recalling information from disc file by backspacing one record at a time and backsubstituting from the last equation upwards until the final three displacements have been found.

The displacements for the nodes of element 3 are then stored in a global vector of nodal displacements, ASDIS.

The eliminated variables associated with element 2 are then reprocessed. In particular the variable for node 5 can be recovered as seen from Stage 4, Fig. 8.12. The disc file information indicates that this was originally stored in equation 4 of the front.

Other nodal displacements for element 2 have already been calculated and are still in the vector of running variables VECRV. It only remains to transfer the displacement of node 5 to the vector ASDIS.

The eliminated variables associated with element 1 are then reprocessed. Disc file information (as well as Stage 1 of Fig. 8.12) shows that it is possible to recover node 1, whose equation is located in the second position in the front.

The displacement for node 1 is added to the global vector of nodal displacements, ASDIS.

Although the elimination has been performed in an untraditional way, the reduced matrix which is transferred to back-up storage, via disc file, can be finally treated in the usual way for backsubstitution. Moving upwards from the last equation, each new equation considered introduces only one unknown quantity which can be directly calculated.

When an element has been reprocessed, all its nodal displacements are known, since *either* they have appeared for the last time with this element in the assembly process and been eliminated and hence recovered on reprocessing of the element, *or* they have appeared later in the assembly process and have already been recovered when the elements are taken in reverse order for backsubstitution.

8.6.2 Case of prescribed displacements

One would like, in cases where nodal displacements are prescribed, to know the magnitude of the concentrated load which must be applied at the node both to balance the force contribution from the neighbouring loaded elements and to produce the imposed displacement, i.e. we wish to find the nodal reaction.

The 4×4 example of Fig. 8.11 is again considered and the reduced form is shown in Fig. 8.14 where:

The equations appear in a different order because of the elimination process.

The primes denote coefficients modified for reduction purposes either in Fig. 8.11 or immediately afterwards.

The displacement δ_3 is prescribed.

$$
\begin{bmatrix}
K_{31} & K_{32} & K_{33} & K_{34} \\
K_{11} & K_{12} & 0 & K_{14} \\
0 & K_{22}' & 0 & K_{24}' \\
0 & 0 & 0 & K_{44}'
\end{bmatrix}
\begin{bmatrix}
\delta_1 \\
\delta_2 \\
\delta_3 \\
\delta_4
\end{bmatrix}
=
\begin{bmatrix}
F_3 + R_3 \\
F_1' \\
F_2' \\
F_4'
\end{bmatrix}
$$

FIG. 8.14.   Backsubstitution stage: reduced matrix when displacement δ_3 is prescribed.

Because of the backsubstitution organisation all other relevant variables are already known (and are still available in the vector of running variables, VECRV) when it comes to determining the unknown associated with equation (3). We can therefore write

$$-R_3 = F_3 - K_{31}\delta_1 - K_{32}\delta_2 - K_{33}\delta_3 - K_{34}\delta_4$$

This appears very much like an ordinary backsubstitution except that the diagonal term, which is usually omitted, is also included in the right hand side summation. The required force is then R_3.

8.6.3 Numerical example

In order to illustrate the backsubstitution phase we will consider the same example used for assembly/elimination: namely Fig. 8.12. The process is displayed in Fig. 8.15.

STAGE	SOLUTION	VECRV
7 and 6 (continued)	$\delta_4 = F_4/K_{44} = 16$	$(\delta_4, 0, 0, 0, 0)$
7 and 6 (continued)	$R_3 = -(F_3 - K_{33}\delta_3 - K_{34}\delta_4) = -90$	$(\delta_4, \delta_3, 0, 0, 0)$
7 and 6	$\delta_2 = (F_2 - K_{23}\delta_3 - K_{24}\delta_4)/K_{22} = 23$	$(\delta_4, \delta_3, \delta_2, 0, 0)$
5 and 4	$\delta_5 = (F_5 - K_{52}\delta_2)/K_{55} = 28$	$(\delta_4, \delta_3, \delta_2, \delta_5, 0)$
3 and 2	$\delta_1 = (F_1 - K_{12}\delta_2 - K_{14}\delta_4)/K_{11} = 24$	$(\delta_4, \delta_1, \delta_2, 0, 0)$

FIG. 8.15.   The backsubstitution stage: Numerical process for the example of Fig. 8.12.

8.7 Equation re-solution facility

For the economic processing of multiple loading cases the inclusion of an equation re-solution facility is essential. In this, the reduced equations are stored in their eliminated form and a second or subsequent solution merely necessitates the reduction of the right hand side load terms. Thus savings are made in two ways. First, the computation of the element stiffnesses is unnecessary and second, the element stiffness assembly and reduction phase is avoided.

An alternative approach is to consider all the load cases at one time by simultaneous reduction of the right hand side load terms. This is economically attractive but does not permit the solution routine to be employed in the iterative solution of non-linear problems. Therefore, in order to provide a more general frontal solver a full re-solution facility is included.

As previously mentioned, the reduced equations and corresponding load terms are stored on disc file. For second and subsequent load cases only the load terms need be modified. Since we assume that disc files can only be accessed sequentially, an additional file must be introduced for storing the reduced loads. (If random access discs are available, use of an extra file can be avoided.) Thus for the first load case, the reduced equation coefficients, housekeeping parameters and load terms are stored on one disc file. For further loading cases, this first file is not modified, but is merely accessed to provide the data required for equation reduction. The reduced load terms for the current load case are then stored on a second disc file.

Assembly of the element stiffnesses is avoided for second and subsequent load cases; a re-solution being identified by the current load case number, ICASE.

8.8 Discussion of the frontal technique

The front solution is a very efficient direct solution process. Its main attraction is that variables are introduced at a later stage and eliminated earlier than in most other methods. The active life of a node lasts from the time in which it first appears in an element to the time in which it last appears in an element. This has the following consequences.

The ordering of elements is crucial and the ordering of nodal numbering is irrelevant. This is an opposite requirement to that for a banded solution and one easier to satisfy since, invariably, there are fewer elements than nodes, especially in three-dimensional problems. Furthermore if a mesh is found to be too coarse in some region, its modification does not require extensive nodal point renumbering.

The core storage requirements are at most the same as those of a banded Gaussian solution. The core storage required is less for structures analysed by means of elements with midside nodes which give "a re-entrant band" in classical schemes. Several examples which justify this statement can be found in references [2] and [5]. One of the most striking is the problem of the closed ring illustrated in Fig. 8.16. With the nodal points numbered

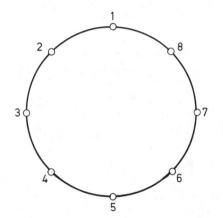
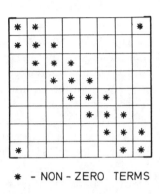

$*$ - NON - ZERO TERMS

FIG. 8.16.   Example of a non-banded equation system.

as shown an ordinary banded solver leads to the half band width being equal to the total number of equations. This is effectively reduced to 3 by the frontal technique. (Of course, the bandwidth for a banded solution can also be reduced by judicious nodal numbering).

Because of the compact nature of the front and because variables are eliminated as soon as conceivably possible, the operations on zero coefficients are minimised and the total arithmetic operations are fewer than with other methods. On the other hand, an elaborate housekeeping system is required for frontal solution. However, since it is entirely performed with integer variables, little storage and computer time is used.

Because any new equation occupies the first available space in the front, there is no need for a bodily shift of the in-core equations as in many other large capacity equation solvers.

8.9 The frontal solution subroutine

8.9.1 Preliminary remarks

The subroutine presented is modular, in that it is a self-contained frontal solver which can be employed in any finite element program. It is assumed that the element stiffness matrices and load vector have been generated elsewhere and are available from disc file or from core storage. It is further assumed that each nodal point has the same number of degrees of freedom; this number being however optional. Although the quadratic isoparametric element is exclusively employed in this text, the frontal solver presented will be compatible with any type of element, subject to the above conditions.

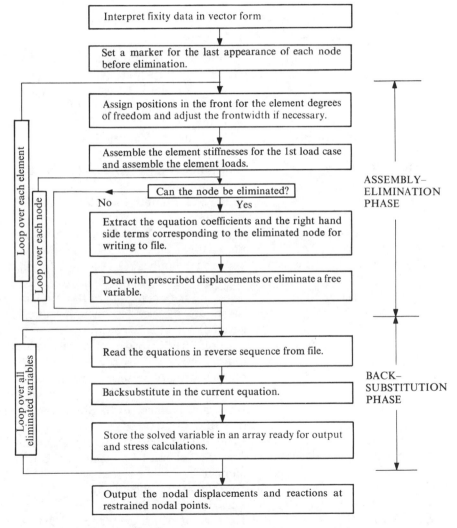

Fig. 8.17. Operation sequence for frontal equation solution.

As subroutine FRONT is long and relatively intricate, a block diagram is presented in Fig. 8.17 showing the main operations performed at each stage. It should be stated that this description is not intended as a flow diagram and for a really detailed description the reader should consult the annotated listing given later in Section 8.9.4.

8.9.2 Dictionary of variable names

Variables are defined in the order in which they first appear in the solution subroutine.

NFUNC(I,J)	A function to compute the position N of element I,J of a square symmetric matrix whose triangular half is stored as a vector
ISTIF,MSTIF	Index and Maximum length of the STIFfness array
IFRON,JFRON,LFRON, NFRON,MFRON	Index, index, index, Number and Maximum FRONtwidth
ITOTV,NTOTV	Index and Number of TOTal Variables in the structure
IPOIN,NPOIN	Index and Number of nodal POINts in the structure
IDOFN,NDOFN	Index and Number of Degrees Of Freedom per Node
IVFIX,NVFIX	Index and Number of FIXed nodal points
NOFIX(IVFIX)	Array of NOdal points which are FIXed
IFPRE(IVFIX,IDOFN)	Array of fixity code values (IF displacements are PREscribed an unit value is entered)
PRESC(IVFIX,IDOFN)	Array of PRESCribed displacement values
IFFIX(ITOTV)	Global vector of fixity codes
FIXED(ITOTV)	Global vector of prescribed displacement values
IELEM,NELEM	Index and Number of ELEMents in the structure
INODE,NNODE	Index and Number of NODes per Element
LNODS(IELEM,INODE)	Array of eLement NODal connection numberS
GSTIF(ISTIF)	The Global STIFfness array
GLOAD(IFRON)	The Global LOAD vector
EQUAT(IFRON)	The array in which the reduced EQUATtions are stored prior to writing to disc

VECRV(IFRON)	The VECtor of Running Variables in which the solved displacements are stored
NACVA(IFRON)	The vector containing a list of the Numbers of the ACtive VAriables in the front
IELVA,KELVA	Index and counter over the number of ELiminated VAriables
IEVAB,JEVAB,KEVAB,NEVAB	Index, index, counter and Number of total Element VAriaBles
ESTIF(IEVAB,JEVAB)	Element STIFfness matrix
LOCEL(IEVAB)	The vector which LOCates the global position of each ELement variable
NIKNO	NIcKname NO. for a particular variable
NDEST(IEVAB)	The DESTination vector for each element
IDEST,JDEST	Working values of the NDEST vector
EQRHS	The Right Hand Side load term after reduction, ready for disc file storage
PIVOT	The diagonal PIVOTing term
ASDIS(NIKNO)	The vector of ASsembled nodal DISplacements

8.9.3 Input/Output diagram

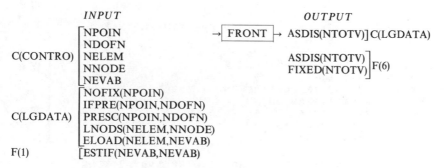

8.9.4 Annotated FORTRAN listing of subroutine FRONT

```
      SUBROUTINE FRONT
      DIMENSION FIXED(160),EQUAT(60),VECRV(160),
     . GLOAD(60),GSTIF(1830),ESTIF(16,16),
     . IFFIX(160),NACVA(60),LOCEL(16),NDEST(16)
```

COMMON BLOCKS

```
      NFUNC(I,J)=(J*J-J)/2+I                               1 *
      MFRON=60                                             2 *
      MSTIF=1830                                           3 *
```

```
C
C*** INTERPRET FIXITY DATA IN VECTOR FORM
C
      NTOTV=NPOIN*NDOFN                                         4 *
      DO 100 ITOTV=1,NTOTV
      IFFIX(ITOTV)=0                                            5 *
  100 FIXED(ITOTV)=0.0
      DO 110 IVFIX=1,NVFIX
      NLOCA=(NOFIX(IVFIX)-1)*NDOFN
      DO 110 IDOFN=1,NDOFN                                      6 *
      NGASH=NLOCA+IDOFN
      IFFIX(NGASH)=IFPRE(IVFIX,IDOFN)
  110 FIXED(NGASH)=PRESC(IVFIX,IDOFN)
C
C*** CHANGE THE SIGN OF THE LAST APPEARANCE
C    OF EACH NODE
C
      DO 140 IPOIN=1,NPOIN                                      7 *
      KLAST=0                                                   8 *
      DO 130 IELEM=1,NELEM                                      9 *
      DO 120 INODE=1,NNODE                                     10 *
      IF(LNODS(IELEM,INODE).NE.IPOIN) GO TO 120
      KLAST=IELEM                                              11 *
      NLAST=INODE
  120 CONTINUE
  130 CONTINUE
      IF(KLAST.NE.0) LNODS(KLAST,NLAST)=-IPOIN                 12 *
  140 CONTINUE                                                 13 *
C
C*** START BY INITIALIZING EVERYTHING THAT
C    MATTERS TO ZERO
C
      DO 150 ISTIF=1,MSTIF                                     14 *
  150 GSTIF(ISTIF)=0.0
      DO 160 IFRON=1,MFRON
      GLOAD(IFRON)=0.0
      EQUAT(IFRON)=0.0                                         15 *
      VECRV(IFRON)=0.0
  160 NACVA(IFRON)=0
C
C*** AND PREPARE FOR DISC READING AND WRITING
C    OPERATIONS
C
      REWIND 1
      REWIND 2                                                 16 *
      REWIND 3
      REWIND 4
C
C*** ENTER MAIN ELEMENT ASSEMBLY-REDUCTION LOOP
C
      NFRON=0                                                  17 *
      KELVA=0                                                  18 *
```

```
            DO 380 IELEM=1,NELEM                          19 *
            KEVAB=0                                        20 *
            READ(1) ESTIF                                  21 *
            DO 170 INODE=1,NNODE                        }  22 *
            DO 170 IDOFN=1,NDOFN
            NPOSI=(INODE-1)*NDOFN+IDOFN                    23 *
            LOCNO=LNODS(IELEM,INODE)                       24 *
            IF(LOCNO.GT.0) LOCEL(NPOSI)=(LOCNO-1)*         25 *
          . NDOFN+IDOFN
            IF(LOCNO.LT.0) LOCEL(NPOSI)=(LOCNO+1)*         26 *
          . NDOFN-IDOFN
        170 CONTINUE                                       27 *
      C
      C*** START BY LOOKING FOR EXISTING DESTINATIONS
      C
            DO 210 IEVAB=1,NEVAB                           28 *
            NIKNO=IABS(LOCEL(IEVAB))                       29 *
            KEXIS=0                                        30 *
            DO 180 IFRON=1,NFRON                          31 *
            IF(NIKNO.NE.NACVA(IFRON)) GO TO 180        }
            KEVAB=KEVAB+1
            KEXIS=1                                     }  32 *
            NDEST(KEVAB)=IFRON
        180 CONTINUE                                       33 *
            IF(KEXIS.NE.0) GO TO 210                       34 *
      C
      C*** WE NOW SEEK NEW EMPTY PLACES FOR
      C    DESTINATION VECTOR
      C
            DO 190 IFRON=1,MFRON                           35 *
            IF(NACVA(IFRON).NE.0) GO TO 190               36 *
            NACVA(IFRON)=NIKNO                          }
            KEVAB=KEVAB+1                                  37 *
            NDEST(KEVAB)=IFRON                          }
            GO TO 200                                      38 *
        190 CONTINUE                                       39 *
      C
      C*** THE NEW PLACES MAY DEMAND AN INCREASE
      C    IN CURRENT FRONTWIDTH
      C
        200 IF(NDEST(KEVAB).GT.NFRON) NFRON=NDEST(KEVAB) 40 *
        210 CONTINUE                                       41 *
      C
      C*** ASSEMBLE ELEMENT LOADS
      C
            DO 240 IEVAB=1,NEVAB                           42 *

            IDEST=NDEST(IEVAB)                             43 *
            GLOAD(IDEST)=GLOAD(IDEST)+ELOAD(IELEM,IEVAB) 44 *
      C
      C*** ASSEMBLE THE ELEMENT STIFFNESSES
      C    - BUT NOT IN RESOLUTION
      C
```

```
      IF(ICASE.GT.1) GO TO 230                           45 *
      DO 220 JEVAB=1,IEVAB                               46 *
      JDEST=NDEST(JEVAB)                                 47 *
      NGASH=NFUNC(IDEST,JDEST)
      NGISH=NFUNC(JDEST,IDEST)
      IF(JDEST.GE.IDEST) GSTIF(NGASH)=                   48 *
    . GSTIF(NGASH)+ESTIF(IEVAB,JEVAB)
      IF(JDEST.LT.IDEST) GSTIF(NGISH)=
    .GSTIF(NGISH)+ESTIF(IEVAB,JEVAB)
  220 CONTINUE
  230 CONTINUE
  240 CONTINUE
C
C*** RE-EXAMINE EACH ELEMENT NODE, TO
C    ENQUIRE WHICH CAN BE ELIMINATED
C
      DO 370 IEVAB=1,NEVAB                               49 *
      NIKNO=-LOCEL(IEVAB)                                50 *
      IF(NIKNO.LE.0) GO TO 370                           51 *
C
C*** FIND POSITIONS OF VARIABLES READY
C    FOR ELIMINATION
C
      DO 350 IFRON=1,NFRON                               52 *
      IF(NACVA(IFRON).NE.NIKNO) GO TO 350               53 *
C
C*** EXTRACT THE COEFFICIENTS OF THE
C    NEW EQUATION FOR ELIMINATION
C
      IF(ICASE.GT.1) GO TO 260                           54 *
      DO 250 JFRON=1,MFRON                               55 *
      IF(IFRON.LT.JFRON) NLOCA=NFUNC(IFRON,JFRON)       56 *
      IF(IFRON.GE.JFRON) NLOCA=NFUNC(JFRON,IFRON)
      EQUAT(JFRON)=GSTIF(NLOCA)                          57 *
  250 GSTIF(NLOCA)=0.0                                   58 *
  260 CONTINUE
C
C*** AND EXTRACT THE CORRESPONDING RIGHT
C    HAND SIDES
C
      EQRHS=GLOAD(IFRON)                                 59 *
      GLOAD(IFRON)=0.0                                   60 *
      KELVA=KELVA+1                                      61 *
C
C*** WRITE EQUATIONS TO DISC OR TO TAPE
C
      IF(ICASE.GT.1) GO TO 270                           62 *
      WRITE(2) EQUAT,EQRHS,IFRON,NIKNO                   63 *
      GO TO 280
  270 WRITE(4) EQRHS                                     64 *
      READ(2) EQUAT,DUMMY,IDUMM,NIKNO
  280 CONTINUE
```

```
C
C*** DEAL WITH PIVOT
C
      PIVOT=EQUAT(IFRON)                                    } 65*
      EQUAT(IFRON)=0.0
C
C*** ENQUIRE WHETHER PRESENT VARIABLE IS
C    FREE OR PRESCRIBED
C
      IF(IFFIX(NIKNO).EQ.0) GO TO 300                         66*
C
C*** DEAL WITH A PRESCRIBED DEFLECTION
C
      ┌──►DO 290 JFRON=1,NFRON
      └ 290 GLOAD(JFRON)=GLOAD(JFRON)-FIXED(NIKNO)*         } 67*
      . EQUAT(JFRON)
      GO TO 340                                               68*
C
C*** ELIMINATE A FREE VARIABLE - DEAL WITH
C    THE RIGHT HAND SIDE FIRST
C
      ┌300 DO 330 JFRON=1,NFRON                               69*
      │    GLOAD(JFRON)=GLOAD(JFRON)-EQUAT(JFRON)*          } 70*
      │    . EQRHS/PIVOT
C
C*** NOW DEAL WITH THE COEFFICIENTS IN CORE
C
      │    IF(ICASE.GT.1) GO TO 320                           71*
      │    IF(EQUAT(JFRON).EQ.0.0) GO TO 330                  72*
      │    NLOCA=NFUNC(0,JFRON)                               73*
      │  ┌─►DO 310 LFRON=1,JFRON
      │  │  NGASH=LFRON+NLOCA
      │  └310 GSTIF(NGASH)=GSTIF(NGASH)-EQUAT(JFRON)*       } 74*
      │    . EQUAT(LFRON)/PIVOT
      │    320 CONTINUE
      └──330 CONTINUE
          340 EQUAT(IFRON)=PIVOT                              75*
C
C*** RECORD THE NEW VACANT SPACE, AND REDUCE
C    FRONTWIDTH IF POSSIBLE
C
      NACVA(IFRON)=0                                          76*
      GO TO 360                                               77*
C
C*** COMPLETE THE ELEMENT LOOP IN THE FORWARD
C    ELIMINATION
C
      └─350 CONTINUE
        360 IF(NACVA(NFRON).NE.0) GO TO 370
            NFRON=NFRON-1                                    } 78*
            IF(NFRON.GT.0) GO TO 360
      └─370 CONTINUE
      └─380 CONTINUE                                         } 79*
```

```
C
C*** ENTER BACK-SUBSTITUTION PHASE, LOOP
C     BACKWARDS THROUGH VARIABLES
C
      DO 410 IELVA=1,KELVA                                    80*
C
C***READ A NEW EQUATION
C
      BACKSPACE 2
      READ(2) EQUAT,EQRHS,IFRON,NIKNO                         }81*
      BACKSPACE 2
      IF(ICASE.EQ.1) GO TO 390
      BACKSPACE 4
      READ(4) EQRHS                                           }82*
      BACKSPACE 4
  390 CONTINUE
C
C*** PREPARE TO BACK-SUBSTITUTE FROM THE
C     CURRENT EQUATION
C
      PIVOT=EQUAT(IFRON)                                      83*
      IF(IFFIX(NIKNO).EQ.1) VECRV(IFRON)=                     84*
    . FIXED(NIKNO)
      IF(IFFIX(NIKNO).EQ.0) EQUAT(IFRON)=0.0                  85*
C
C*** BACK-SUBSTITUTE IN THE CURRENT EQUATION
C
      DO 400 JFRON=1,MFRON                                    }86*
  400 EQRHS=EQRHS-VECRV(JFRON)*EQUAT(JFRON)
C
C*** PUT THE FINAL VALUES WHERE THEY BELONG
C
      IF(IFFIX(NIKNO).EQ.0) VECRV(IFRON)=                     }87*
    . EQRHS/PIVOT
      IF(IFFIX(NIKNO).EQ.1) FIXED(NIKNO)=-EQRHS               88*
      ASDIS(NIKNO)=VECRV(IFRON)                               89*
  410 CONTINUE                                                90*
      WRITE(6,900)
  900 FORMAT(1H0,5X,13HDISPLACEMENTS)
      IF(NDOFN.NE.2) GO TO 430
      IF(NDIME.NE.1) GO TO 420
      WRITE(6,905)
  905 FORMAT(1H0,5X,4HNODE,6X,5HDISP.,7X,
    . 8HROTATION)
      GO TO 440                                               }91*
  420 WRITE(6,910)
  910 FORMAT(1H0,5X,4HNODE,5X,7HX-DISP.,
    . 7X,7HY-DISP.)
      GO TO 440
  430 WRITE(6,915)
  915 FORMAT(1H0,5X,4HNODE,6X,5HDISP.,8X,
    . 7HXZ-ROT.,7X,7HYZ-ROT.)
  440 CONTINUE
```

```
        ┌──────▶ DO 450 IPOIN=1,NPOIN
        │        NGASH=IPOIN*NDOFN
        │        NGISH=NGASH-NDOFN+1                                    ⎫
        └────450 WRITE(6,920) IPOIN,(ASDIS(IGASH),IGASH=               ⎬92*
               . NGISH,NGASH)                                           ⎭
           920 FORMAT(I10,3E14.6)
               WRITE(6,925)
           925 FORMAT(1H0,5X,9HREACTIONS)                              ⎫
               IF(NDOFN.NE.2) GO TO 470                                 ⎪
               IF(NDIME.NE.1) GO TO 460                                 ⎪
               WRITE(6,930)                                             ⎪
           930 FORMAT(1H0,5X,4HNODE,6X,5HFORCE,8X,6HMOMENT)            ⎪
               GO TO 480                                                ⎪
           460 WRITE(6,935)                                            ⎬93*
           935 FORMAT(1H0,5X,4HNODE,5X,7HX-FORCE,7X,                    ⎪
               . 7HY-FORCE)                                            ⎪
               GO TO 480                                                ⎪
           470 WRITE(6,940)                                            ⎪
           940 FORMAT(1H0,5X,4HNODE,6X,5HFORCE,6X,                      ⎪
               . 9HXZ-MOMENT,5X,9HYZ-MOMENT)                           ⎭
           480 CONTINUE
        ┌──────▶ DO 510 IPOIN=1,NPOIN
        │        NLOCA=(IPOIN-1)*NDOFN
        │  ┌───▶ DO 490 IDOFN=1,NDOFN
        │  │     NGUSH=NLOCA+IDOFN
        │  │     IF(IFFIX(NGUSH).GT.0) GO TO 500
        │  └─490 CONTINUE                                              ⎫
        │        GO TO 510                                             ⎬94*
        │    500 NGASH=NLOCA+NDOFN                                      ⎪
        │        NGISH=NLOCA+1                                         ⎪
        │        WRITE(6,945) IPOIN,(FIXED(IGASH),IGASH=               ⎪
        │        . NGISH,NGASH)                                        ⎪
        └──510 CONTINUE                                                ⎪
           945 FORMAT(I10,3E14.6)                                      ⎭
       C
       C*** POST FRONT - RESET ALL ELEMENT CONNECTION
       C    NUMBERS TO POSITIVE VALUES FOR SUBSEQUENT
       C    USE IN STRESS CALCULATION
       C
        ┌──────▶ DO 520 IELEM=1,NELEM                                   ⎫
        │  ┌───▶ DO 520 INODE=1,NNODE                                  ⎬95*
        └──520 LNODS(IELEM,INODE)=IABS(LNODS(IELEM,INODE))            ⎭
               RETURN
               END
```

1* Set up the function giving the address of stiffness term K_{ij} in GSTIF. This is valid for $i \leqslant j$

2* Define the maximum permissible frontwidth

3* Define the maximum number of positions allowed in the stiffness vector

4* Determine the total number of degrees of freedom in the structure

5* Initialise the vectors which will contain the fixity codes and prescribed displacement values

6* Transfer the fixity codes and prescribed displacement values from the two-dimensional arrays, input in subroutine INPUT, to vectors ranging over the total number of degrees of freedom in the structure

7* Enter loop over each node

8* Set counter for element number to zero

9* Enter loop over each element

10* Enter loop over each node within an element

11* If an element node number coincides with the node number under consideration in the nodal point loop, record the element number and the nodal position within the element. Since a node should appear only once in an element this coincidence should, at most, occur once only per element. The search is conducted in the same order of increasing element numbering to be employed in frontal solution, the last found value only being recorded by overwriting

12* After searching through each element, the last appearance of a node in an element is recorded with a negative sign in the LNODS array

13* Loop to begin the process for the next node number

14* Initialise the stiffness vector to zero

15* Initialise the load vector, reduced equations vector, the vector of solved variables and the active variables array respectively

16* Rewind all disc files

17* Set the current frontwidth to zero

18* Set the counter over all the eliminated variables in the structure to zero

19* Enter the main loop over each element

20* Set the counter over the variables per element to zero

21* Read the stiffness matrix of the element from disc file

22* Enter loops over the element nodes and degrees of freedom per node respectively

23* Locate, on an element level, the position of a particular degree of freedom of a particular node of the element under consideration

24* Determine the nodal number of the element node under consideration

25* Compute the global location of the element degree of freedom under consideration and store in the local element array, LOCEL

26* If it is a last appearance (identified by a negative value of LOCNO) place a negative sign before the relevant entry in LOCEL

27* Loop to consider the next variable (degree of freedom)

28* Enter loop over each variable of an element

29* Compute the global location of the variable

30* Set counter to be used in identification of existing destination locations to zero

31* Enter loop over each variable in the front

. 32* If the variable under consideration, NIKNO, already exists in the front, increment the counter over the total degrees of freedom per element, KEVAB, by one, set KEXIS to one to indicate that a destination location for this variable already exists and store the frontal location, IFRON, as the variable destination location in the destination vector NDEST

33* Loop to continue search over the remaining variables in the front

34* If KEXIS = 1, indicating that a destination location for variable NIKNO already exists, go to 210 and omit the section which seeks empty spaces in the front

35* Enter loop over every permissible location in the front

36* If the frontal location is full (being non-zero) loop to next location

37* If the frontal position is empty, store variable NIKNO in this position and register this fact in NACVA. Also increment the counter over the total degrees of freedom per element, KEVAB, by one and store the frontal location, IFRON, being the variable destination location, in the destination vector IDEST

38* As soon as a location in front has been found for variable NIKNO, discontinue the search

39* If the frontal location was already full, continue the search over the remaining positions

40* If, in order to accommodate the variable NIKNO, the variable has had to be located at the end of the present front, then increase the frontwidth to include this

41* Loop to locate the remaining element variables in the front

42* Loop over each element variable to give the row location i in K_{ij}.

43* Set IDEST equal to the destination vector of the variable under consideration, i.e. IDEST defines the position in the global stiffness and load vectors

44* Insert the element loads in the correct position in the global load vector

45* If the current load case is not the first, avoid assembling the element stiffnesses

46* Loop over each element variable, to give the column location j in K_{ij}

47* Set JDEST equal to the destination vector of the variable under consideration in this inner loop, i.e. define the global column location

48* Store the element stiffnesses in the correct locations in the one-dimensional global stiffness submatrix. The function NFUNC is used for this purpose with provision taken if i is less than or greater than j.

49* Enter loop over each element variable to check which ones can be eliminated

50* Store the variable number as NIKNO with a change of sign. Therefore for elimination we are only interested in a *positive* value of NIKNO

51* If NIKNO is not positive, loop to the next variable

52* Enter loop over each variable in the front to find the position of variable to the eliminated

53* If the current front variable is not the one to be eliminated, NIKNO, go to the end of the frontal variable loop and continue the search

54* If the current load case is not the first, avoid extracting the coefficients of the equation about to be eliminated.

55* Enter loop for extracting the coefficients of the new equation for elimination. Note here that we loop over the complete front, not merely the current frontwidth, so that all the positions in array EQUAT are re-defined and hence need not be initialised

56* Compute the location of the eliminated variable in the global stiffness submatrix

57* Extract the appropriate stiffness coefficient and insert into the equation to be eliminated which will later be transferred to backing disc store

58* To make its space available for subsequent use we replace the stiffness coefficient by zero

59* Extract the corresponding load term from the load subvector ready to be transferred to disc file

60* Replace the extracted term by zero in the load vector

61* Increase the counter of eliminated variables by one to keep a running total of the eliminated variables

62* If the current load case is not the first, do not write information relating to the eliminated equation to disc as this information already exists on file

63* Write the equation and reduced load term corresponding to the eliminated variable to disc file. Also store the current position of this variable in the front, IFRON, and the variable number, NIKNO

64* For the second and subsequent load cases write the reduced load term to a separate disc file. Also read one record from file 2; since this information is required for reducing the remaining equations. Note that the load term is read as a dummy value to avoid overwriting the current term by the value corresponding to the first loading case. Also the frontal position is read as a dummy value, IDUMM, to avoid changing the parameter of DO LOOP 350.

65* Store the diagonal term of the eliminated equation as PIVOT and replace by zero in the corresponding equation position as a temporary convenience

66* If there is no prescribed displacement associated with the variable, go to 300 and skip the fixity computation

67* This prescribed variable process was described in Section 8.5.3. and Fig. 8.11. Only the right hand side of the equations need be modified as indicated

68* For the case of a prescribed variable the normal elimination procedure is omitted

69* Loop over each frontal variable. This is the most expensive loop of the frontal process

70* Reduce the right hand side load terms as indicated in Section 8.5.2 and Fig. 8.10.

71* For the second and subsequent load cases avoid reducing the stiffness terms

72* This avoids time-wasting on zero operations

73* This statement removes the function evaluation of the global stiffness submatrix positions from the innermost expensive loop

74* Perform the submatrix elimination as indicated in Section 8.5.2 and Fig. 8.10. Note that the upper DO LOOP limit is JFRON and not NFRON, since we have to process GSTIF which is an upper triangle

75* The pivot term must now be replaced in EQUAT as the reduced equation has not yet been written to disc

76* The vacant space in front associated with the eliminated variable is recorded via NACVA

77* Having eliminated a variable leave the loop over the frontal locations (since the variable should not appear again in the front) and process the next element variable

78* The current front length, NFRON, should just span the non-zero entries in NACVA. If the last entry in NACVA has just been made zero in 76* it is possible that a sequence of consecutive zeroes have been exposed at the end of the front. Thus the frontwidth is compressed until a non-zero entry is found or until the frontwidth is zero

79* Termination of loops over the element variables and elements respectively. At this stage the assembly-elimination process is complete

80* Enter the backsubstitution phase by looping over the total number of eliminated variables

81* Recall a new equation corresponding to an eliminated variable from disc file. The disc file was left in its final position after the assembly–elimination process. Thus we backspace one record, read the

information and backspace again so as to be ready to repeat the
sequence for the next equation.

82* For the second and subsequent load cases overwrite the load term
read from disc file 2 (which is the value corresponding to the first
loading case) by the current value as read from disc file 4.

83* As in the elimination process, it is convenient to extract the pivot to a
working location

84* If the variable has a prescribed value, store this value in the running
vector of solved variables, VECRV

85* If the variable is free, set the corresponding diagonal position to zero.
Since the diagonal term must not multiply by any spurious value
that has been left in the corresponding position in the vector of running variables, VECRV

86* Calculate the right hand side term minus the left hand side contributions i.e. calculate the residual

87* If the variable is free, calculate the variable displacement by dividing
the residual by the pivot term and store in VECRV

88* If the variable has a prescribed value, the reaction value is computed
as the negative of the residual as discussed in Section 8.6.2 and is
stored in the corresponding location in the vector of prescribed displacements, FIXED

89* Insert the variable displacement in the global displacement vector,
ASDIS

90* Loop to process the next variable

91* Write headings for the displacement output depending on the application being considered. The number of degrees of freedom per node,
NDOFN and the number of coordinate dimensions, NDIME are
used to identify the type of problem being solved.

92* Output the nodal displacements

93* Write headings for the reactions at nodal points having prescribed
displacements, depending on the application under consideration

94* Output the reactions for nodal variables having prescribed displacements; these being located by use of the fixity code vector, IFFIX.

95* Reset all the element connection numbers, LNODS, to positive values.
This is necessary for use in stress evaluation and for the regeneration of
element loads for subsequent loading cases

The diagonal term of an eliminated equation is stored as PIVOT in Step 65*.
This value should be checked to ensure that it is positive since firstly, this is a
physical requirement for structural problems, and also since later in the
equation reduction process terms will be divided by it. Other diagnostic
checks for roundoff problems can also be inserted [6].

8.9.5 The dimensioned arrays in subroutine FRONT

The arrays which are local to subroutine FRONT are dimensioned by use of a DIMENSION statement and do not appear in any COMMON block. The required array dimensions are indicated below where the DIMENSION statement is listed in terms of program variables

```
DIMENSION  FIXED(MPOIN*MDOFN), EQUAT(MFRON),
           VECRV(MPOIN*MDOFN), GLOAD(MFRON),
           GSTIF(MSTIF), ESTIF(MEVAB,MEVAB),
           IFFIX(MPOIN*MDOFN), NACVA(MFRON),
           LOCEL(MEVAB), NDEST(MEVAB)
```

The indices employed are described below

MPOIN   The maximum number of nodal points for which the program is to be dimensioned

MDOFN   The maximum number of degrees of freedom per node for which the program is to be dimensioned

MEVAB   The maximum number of nodal variables per element for which the program is to be dimensioned

MFRON   The maximum permissible number of variables (degrees of freedom) allowed in the front

MSTIF   The maximum permissible number of positions in the one-dimensional global stiffness array

Arrays EQUAT, GLOAD, GSTIF and NACVA must be dimensioned according to the values of MSTIF and MFRON defined in the 2nd and 3rd statements of the subroutine respectively. Any dimension changes made in the program, for the accommodation of larger problems, must be accompanied by appropriate changes in these two statements.

It should be noted that the maximum frontwidth, MFRON, cannot be calculated from a knowledge of the other program parameters, such as the maximum number of nodes, MPOIN, the maximum number of elements, MELEM, etc. The maximum frontwidth MFRON will depend on the structural geometry as well as the order of element numbering. However the maximum space required in the global stiffness array, GSTIF, can be calculated for a given maximum frontwidth, from the expression

$$MSTIF = (MFRON*MFRON - MFRON)/2 + MFRON$$

$$= MFRON*(MFRON + 1)/2.$$

So that for a maximum permissible frontwidth, MFRON = 60, the dimension of MSTIF must be 1830 which is the value specified in the program.

8.10 Summary

In view of the complexity of this chapter, the basic steps of the frontal solution technique will now be summarised and its basic advantages reiterated.

Equation assembly and reduction proceeds element by element. This requires a relatively elaborate "housekeeping" technique.

Each element variable is assigned a location in the front and the frontwidth adjusted if necessary.

The element stiffness and load terms are assembled into the global equations as governed by the position of the variable in the front.

If a node is appearing for the last time the corresponding degrees of freedom are eliminated by equation reduction. The stiffness and load terms corresponding to the eliminated variables are stored on disc file.

The displacements are obtained by backsubstitution in reverse order.

The main advantages of the frontal process are:

Nodal numbering is irrelevant and it is the order in which elements are numbered that is of prime importance. This numbering condition is well suited to isoparametric elements, since the number of elements is always far less than the number of nodal points for such complex elements.

Variables are introduced at a later stage and eliminated earlier than in most methods. This helps to reduce numerical inaccuracies due to roundoff.

References

1. Ralston, A., "A First Course in Numerical Analysis". McGraw-Hill, New York, 1965.
2. Irons, B. M., A frontal solution program. *Int. J. Num. Meth. Eng.* **2**, 5–32, 1970.
3. Kan, D. K. Y., "A Simple Front Solution Program for Finite Element Techniques." Report CNME/CR/51, Dept. of Civ. Eng., University College of Swansea, 1971.
4. Irons, B. M., and Fawkes, A., "Students' Front Solution for Varying Degrees of Freedom per Node". Internal Report, Dept. of Civil Eng., University College of Swansea, 1973.
5. Melosh, R. J., and Bamford, R. M., Efficient solution of load-deflection equations. *J. Struct. Div. ASCE*, **95**, 661–676, 1965.
6. Irons, B. M., Roundoff criteria in direct stiffness solutions. *AIAA J.*, **6**, 1308–1312, 1968.

9

Data Checking and Error Diagnostics

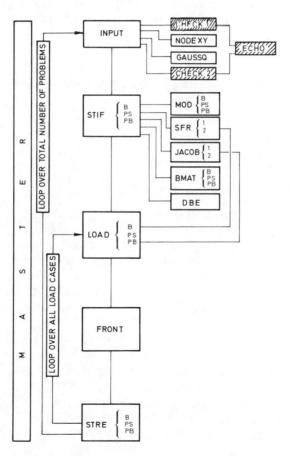

9.1 Introduction

The principle of including in a program ample diagnostics—those printed messages which tell the user what is wrong with his data and why the machine has stopped—has already been almost universally accepted by software houses. The quest for dependable diagnostics leads to one of several relatively sophisticated housekeeping techniques, the inclusion of which unfortunately inevitably distracts from the essential operations of the finite element method. For this reason error diagnostics have here been separated into three additional subroutines which will be described in this chapter. A more elaborate diagnostic system is described in Reference [1].

The three error diagnostic subroutines described are employed to scrutinise the data entered in subroutine INPUT. As soon as the data card containing the problem control parameters has been input, subroutine CHECK1 is called from subroutine INPUT to check these variables. If any errors are detected subroutine ECHO is called by CHECK1 to echo the remainder of the problem input data via the lineprinter before the job is terminated. If no errors are detected in the control parameters, then the geometric data, boundary cond'tions and material properties are assimilated by subroutine INPUT and subroutine CHECK 2 is finally called to scrutinise this information. Again if errors are detected, the remaining data is echoed by subroutine ECHO before termination of the computer run.

9.2 Error diagnostic subroutine CHECK1

The function of this subroutine is to scrutinise the problem control parameters, which are contained on the first data card read in subroutine INPUT. Since subroutine INPUT is common to all three applications (i.e. beam, plane stress/strain and plate bending) subroutine CHECK1 will only check that the control parameters are within the bounds defined by the correct values for the three cases.

A counter, KEROR, is employed in the subroutine to indicate whether or not any errors have been detected. If errors have been found (indicated by KEROR = 1) subroutine ECHO, described in the next section is called to list the remainder of the input data.

Any errors detected are signalled by means of printed error numbers. The interpretation of each error message is given in Section 9.5.

Subroutine CHECK1 is now listed and descriptive notes provided.

Input/Output diagram

	INPUT		OUTPUT

```
        ⎡ NPOIN  →  ┌─────────┐ → NEROR(IEROR)] C(WORK)
        │ NELEM     │ CHECK1  │
        │ NNODE     └─────────┘
        │ NVFIX
        │ NCASE
        │ NTYPE
C(CONTRO)│ NDOFN
        │ NMATS
        │ NPROP
        │ NGAUS
        │ NDIME
        ⎣ NSTRE
```

Annotated FORTRAN

```
      SUBROUTINE CHECK1
C
C*** TO CRITICIZE THE DATA CONTROL CARD AND
C    PRINT ANY DIAGNOSTICS
C
```

┌─────────────────────┐
│   COMMON BLOCKS   │
└─────────────────────┘

```
┌──► DO 10 IEROR=1,24                                      ⎫ 1*
└── 10 NEROR(IEROR)=0                                      ⎭
C
C*** CREATE THE DIAGNOSTIC MESSAGES
C
      IF(NPOIN.LE.0) NEROR(1)=1                               2*
      IF(NELEM*NNODE.LT.NPOIN) NEROR(2)=1                     3*
      IF(NVFIX.LT.1.OR.NVFIX.GT.NPOIN) NEROR(3)=1            4*
      IF(NCASE.LE.0) NEROR(4)=1                               5*
      IF(NTYPE.LT.0.OR.NTYPE.GT.2) NEROR(5)=1               6*
      IF(NNODE.LT.3.OR.NNODE.GT.8) NEROR(6)=1               7*
      IF(NDOFN.LT.2.OR.NDOFN.GT.3) NEROR(7)=1               8*
      IF(NMATS.LE.0.OR.NMATS.GT.NELEM) NEROR(8)=1           9*
      IF(NPROP.LT.3.OR.NPROP.GT.5) NEROR(9)=1              10*
      IF(NGAUS.LT.2.OR.NGAUS.GT.3) NEROR(10)=1             11*
      IF(NDIME.LT.1.OR.NDIME.GT.2) NEROR(11)=1             12*
      IF(NSTRE.LT.2.OR.NSTRE.GT.5) NEROR(12)=1             13*
C
C*** EITHER RETURN,OR ELSE PRINT THE ERRORS
C    DIAGNOSED
C
      KEROR=0                                                14*
┌──► DO 20 IEROR=1,12                                       15*
│     IF(NEROR(IEROR).EQ.0) GO TO 20                         16*
```

```
        KEROR=1
        WRITE(6,900) IEROR
 900 FORMAT(//25H *** DIAGNOSIS BY CHECK1,         } 17*
    . 6H ERROR,I3)
  20 CONTINUE
        IF(KEROR.EQ.0) RETURN                             18*
C
C*** OTHERWISE ECHO ALL THE REMAINING DATA
C    WITHOUT FURTHER COMMENT
C

        CALL ECHO                                         19*
        END
```

1* Set the error indicators for all 24 error types which the diagnostic subroutines will detect to zero.

2* If the specified total number of nodal points, NPOIN, in the structure is less than or equal to zero set the indicator for error type 1 equal to 1.

3* If the possible maximum total number of nodal points in the structure (number of elements × number of nodes per element) is less than the specified total number of nodal points in the structure, set the indicator for error type 2 equal to 1.

4* If the number of restrained nodal points is less than 1 or greater than the total number of nodal points, NPOIN, set the indicator for error type 3 equal to 1. At least one point of the structure must be restrained in order to prevent rigid body motions. (In fact for plane applications, at least 2 nodal points must be restrained.)

5* If the total number of loading cases is less than or equal to zero set the indicator for error type 4 equal to 1.

6* If the problem type parameter for plane situations, NTYPE, is less than zero (beam or plate problems), or greater than 2 (plane strain) set the indicator for error type 5 equal to 1.

7* If the number of nodes per element is less than 3 (beam) or greater than 8 (plane or plate problems) set the indicator for error type 6 equal to 1.

8* If the number of degrees of freedom per node is less than 2 (beam or plane problems) or greater than 3 (plate bending situations) set the indicator for error type 7 equal to 1.

9* If the total number of different materials is less than or equal to zero or greater than the total number of elements in the structure set the indicator for error type 8 equal to 1.

10* If the number of individual properties per material is less than 3 (beam), or greater than 5 (plane stress/strain) set the indicator for error type 9 equal to 1.

11* If the order of Gaussian numerical integration is not specified as either 2 or 3 set the indicator for error type 10 equal to 1.

12* If the number of coordinate dimensions is not equal to either 1 (beam) or 2 (plane and plate problems) set the indicator for error type 11 equal to 1.

13* If the number of generalised stress components is less than 2 (beam), or greater than 5 (plate bending situations) set the indicator for error type 12 equal to 1.

14* Initialise the indicator which will be used to signal whether or not any errors have been detected.

15* Loop over each error type.

16* If the error indicator for a particular error type is zero, go to the end of the loop and process the next error type.

17* Otherwise set the error counter, KEROR, equal to 1 and output the error message.

18* If no errors have been detected by this subroutine, return to subroutine INPUT.

19* Otherwise call subroutine ECHO to list the remainder of the input data.

9.3 Data echo subroutine, ECHO

The function of this subroutine is to list all the remaining data cards after at least one error has been detected by either of the diagnostic subroutines CHECK1 or CHECK2. This is accomplished by means of a simple read and write operation in alphanumeric format. The program will always fail in this subroutine by attempting to read beyond the end of the input data file.

Input/Output diagram

$$\text{INPUT} \qquad\qquad\qquad \text{OUTPUT}$$

$$\text{F(5) [NTITL(80)} \rightarrow \boxed{\text{ECHO}} \rightarrow \text{NTITL(80)] F(6)}$$

Annotated FORTRAN

```
      SUBROUTINE ECHO
      DIMENSION NTITL(80)
```

COMMON BLOCKS

```
      WRITE(6,900)
  900 FORMAT(//25H NOW FOLLOWS A LISTING OF,      }1*
     . 25H POST-DISASTER DATA CARDS/)
   10 READ(5,905) NTITL                           }2*
  905 FORMAT(80A1)
```

```
      WRITE(6,910) NTITL                          }3*
  910 FORMAT(20X,80A1)
      GO TO 10                                      4*
      END
```

1* Write a heading to signal the beginning of the data echo.
2* Read a card in alphanumeric format.
3* Write, in alphanumeric format, the information contained on the card.
4* Return to repeat the operation for the next card.

9.4 Error diagnostic subroutine CHECK2

If it has been ascertained, in subroutine CHECK1, that the problem control
parameters have not been assigned impossible values, the geometric data,
boundary conditions and material properties are then assimilated by subroutine INPUT. This data is then scrutinised for possible errors in subroutine
CHECK2 where error types 13 to 24 are checked for. The errors which are
diagnosed can be seen from the list of error messages given in Section 9.5.

Probably the most useful check in this subroutine is the one which ensures
that the maximum frontwidth does not exceed the value specified in subroutine FRONT. Subroutine CHECK2 is now listed and explanatory notes
provided.

Input/Output diagram

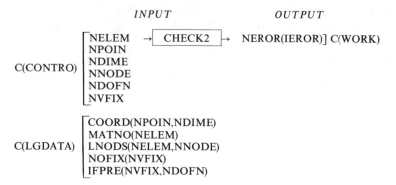

Annotated FORTRAN Listing

```
      SUBROUTINE CHECK2
C
C*** TO CRITICIZE THE DATA FROM SUBROUTINE INPUT
C
```

```
      DIMENSION NDFRO(25)
```

COMMON BLOCKS

```
      MFRON=60                                              1 *
C
C*** CHECK AGAINST TWO IDENTICAL NONZERO
C    NODAL COORDINATES
C
      DO 10 IELEM=1,NELEM                                   } 2 *
   10 NDFRO(IELEM)=0
      DO 40 IPOIN=2,NPOIN                                   3 *
      KPOIN=IPOIN-1                                         } 4 *
      DO 30 JPOIN=1,KPOIN
      DO 20 IDIME=1,NDIME
      IF(COORD(IPOIN,IDIME).NE.COORD(JPOIN,                 } 5 *
     . IDIME)) GO TO 30
   20 CONTINUE
      NEROR(13)=NEROR(13)+1                                 6 *
   30 CONTINUE
   40 CONTINUE
C
C*** CHECK THE LIST OF ELEMENT PROPERTY NUMBERS
C
      DO 50 IELEM=1,NELEM
   50 IF(MATNO(IELEM).LE.0.OR.MATNO(IELEM).GT.              } 7 *
     . NMATS) NEROR(14)=NEROR(14)+1
C
C*** CHECK FOR IMPOSSIBLE NODE NUMBERS
C
      DO 70 IELEM=1,NELEM                                   8 *
      DO 60 INODE=1,NNODE                                   9 *
      IF(LNODS(IELEM,INODE).EQ.0) NEROR(15)=                } 10 *
     . NEROR(15)+1
   60 IF(LNODS(IELEM,INODE).LT.0.OR.LNODS(IELEM,            } 11 *
     . INODE).GT.NPOIN) NEROR(16)=NEROR(16)+1
   70 CONTINUE
C
C*** CHECK FOR ANY REPETITION OF A NODE
C    NUMBER WITHIN AN ELEMENT
C
      DO 140 IPOIN=1,NPOIN                                  } 12 *
      KSTAR=0
      DO 100 IELEM=1,NELEM                                  } 13 *
      KZERO=0
      DO 90 INODE=1,NNODE                                   } 14 *
      IF(LNODS(IELEM,INODE).NE.IPOIN) GO TO 90

      KZERO=KZERO+1                                         15 *
      IF(KZERO.GT.1) NEROR(17)=NEROR(17)+1                  16 *
C
C*** SEEK FIRST,LAST AND INTERMEDIATE
C    APPEARANCES OF NODE IPOIN
C
```

```
        IF(KSTAR.NE.0) GO TO 80                              17*
        KSTAR=IELEM                                          18*
C
C*** CALCULATE INCREASE OR DECREASE IN
C    FRONTWIDTH AT EACH ELEMENT STAGE
C
        NDFRO(IELEM)=NDFRO(IELEM)+NDOFN                      19*
     80 CONTINUE
C
C*** AND CHANGE THE SIGN OF THE LAST
C    APPEARANCE OF EACH NODE
C
        KLAST=IELEM                                          20*
        NLAST=INODE                                          21*
     90 CONTINUE
    100 CONTINUE
        IF(KSTAR.EQ.0) GO TO 110                             22*
        IF(KLAST.LT.NELEM) NDFRO(KLAST+1)=             ⎫
      . NDFRO(KLAST+1)-NDOFN                           ⎬23*
        LNODS(KLAST,NLAST)==IPOIN                            24*
        GO TO 140                                           25*
C
C*** CHECK THAT COORDINATES FOR AN UNUSED
C    NODE HAVE NOT BEEN SPECIFIED
C
    110 WRITE(6,900) IPOIN                             ⎫
    900 FORMAT(/15H CHECK WHY NODE,I4,                 ⎬26*
      . 14H NEVER APPEARS)                             ⎭
        NEROR(18)=NEROR(18)+1                               27*
        SIGMA=0.0                                      ⎫
        DO 120 IDIME=1,NDIME                           ⎬28*
    120 SIGMA=SIGMA+ABS(COORD(IPOIN,IDIME))            ⎭
        IF(SIGMA.NE.0.0) NEROR(19)=NEROR(19)+1              29*
C
C*** CHECK THAT AN UNUSED NODE NUMBER IS NOT
C    A RESTRAINED NODE
C
        DO 130 IVFIX=1,NVFIX                           ⎫
    130 IF(NOFIX(IVFIX).EQ.IPOIN) NEROR(20)=           ⎬30*
      . NEROR(20)+1                                    ⎭
    140 CONTINUE
C
C*** CALCULATE THE LARGEST FRONTWIDTH
C
        NFRON=0                                        ⎫
        KFRON=0                                        ⎬31*
        DO 150 IELEM=1,NELEM                                32*
        NFRON=NFRON+NDFRO(IELEM)                            33*
    150 IF(NFRON.GT.KFRON) KFRON=NFRON                      34*
        WRITE(6,905) KFRON                             ⎫
    905 FORMAT(//28HMAX FRONTWIDTH ENCOUNTERED =,I5) ⎬35*
        IF(KFRON.GT.MFRON) NEROR(21)=1                      36*
```

```
C
C*** CONTINUE CHECKING THE DATA FOR THE
C    FIXED VALUES
C
      DO 170 IVFIX=1,NVFIX                                       37*
      IF(NOFIX(IVFIX).LE.0.OR.NOFIX(IVFIX).
     . GT.NPOIN) NEROR(22)=NEROR(22)+1                           38*
      KOUNT=0                                                    39*
      DO 160 IDOFN=1,NDOFN
  160 IF(IFPRE(IVFIX,IDOFN).GT.0) KOUNT=1                        40*
      IF(KOUNT.EQ.0) NEROR(23)=NEROR(23)+1                       41*
      KVFIX=IVFIX-1
      DO 170 JVFIX=1,KVFIX                                       42*
  170 IF(IVFIX.NE.1.AND.NOFIX(IVFIX).EQ.
     . NOFIX(JVFIX)) NEROR(24)=NEROR(24)+1                       43*
      KEROR=0                                                    44*
      DO 180 IEROR=13,24                                         45*
      IF(NEROR(IEROR).EQ.0) GO TO 180                            46*
      KEROR=1
      WRITE(6,910) IEROR,NEROR(IEROR)
  910 FORMAT(//30H*** DIAGNOSIS BY CHECK2, ERROR,                47*
     . I3,6X,18H ASSOCIATED NUMBER,I5)
  180 CONTINUE
      IF(KEROR.NE.0) GO TO 200                                   48*
C
C*** RETURN ALL NODAL CONNECTION NUMBERS TO
C    POSITIVE VALUES
C
      DO 190 IELEM=1,NELEM
      DO 190 INODE=1,NNODE                                       49*
  190 LNODS(IELEM.INODE)=IABS(LNODS(IELEM,INODE))
      RETURN
  200 CALL ECHO                                                  50*
      END
```

1* Set the maximum possible frontwidth, MFRON, equal to the same value defined in subroutine FRONT.

2* Set the NDFRO array to zero. This will store the contribution to the frontwidth of each element.

3* Enter loop over all nodal points, except the first.

4* Enter loop over the nodal points up to, but not including, the node being currently checked.

5* Check if the coordinates of the current node are equal to those of one or more of the preceding nodes. If not, go to 30, thereby bypassing the error counter.

6* Every time the coordinates of the node being currently considered are equal to those of the previous node, increase the error counter for error type 13 by 1.

7* Check for each element that the material property identification number is not less than or equal to zero or greater than the total number of materials previously specified as a control parameter. For every violation increase the error counter for error type 14 by 1.

8* Enter loop over all the elements.

9* Enter loop over all the nodes of an element.

10* If any element nodal connection number is zero, increment the error counter for error type 15 by 1.

11* If any element nodal connection number is negative or greater than the specified total number of nodal points in the structure, NPOIN, increase the error counter for error type 16 by 1.

12* Enter loop over each nodal point in the structure and initialise, KSTAR, to zero. KSTAR will be used to denote the element in which the node currently being considered first appears.

13* Enter loop over each element and initialise KZERO to zero. KZERO will be used to indicate how many times a particular node appears in any element.

14* Enter loop over each nodal point of an element and if the nodal connection number is not equal to the nodal point being currently considered loop to the next nodal connection number.

15* Increment KZERO by 1 to indicate that the current node has been found in the element.

16* If the current node appears more than once in the element, increase the error counter for error type 17 by 1.

17* If KSTAR is not equal to zero go to 80 and avoid the frontwidth calculation.

18* Set KSTAR equal to the element number in which the current node has been first found.

19* Add the contribution to the frontwidth on assembly of the current node.

20* Each appearance of a node (including the first) could be the last. Thus KLAST is continually updated to be the element in which the node was last detected.

21* Similarly modify NLAST for the nodal connection number within the element.

22* If KSTAR = 0, indicating that a node number has not been found in the list of element nodal connection numbers, branch to 110 to print an error message.

23* After the last appearance of a node reduce the frontwidth appropriately.

24* Each last appearance, is labelled by introducing a negative sign before the nodal connection number in the LNODS array.

25* Continue the search to the next node number.

26* If a particular node number does not appear in the list of nodal connection numbers, print a warning message.

27* And increase the error counter for error type 18 by 1.

28* For a node that does not appear in the list of nodal connection numbers, sum the absolute values of its nodal coordinates.

29* If the sum of the absolute values of the nodal coordinates are non-zero, increase the error counter for error type 19 by 1; since it is illogical that coordinates have been supplied for a node, which are then not used.

30* If an unused node appears in the list of restrained nodal points, increase the error counter for error type 20 by 1.

31* Initialise to zero the current frontwidth, NFRON, and the maximum encountered to date, KFRON.

32* Enter loop over each element.

33* Increase the frontwidth with the first appearance of each node and decrease the frontwidth with each last appearance.

34* Record the instantaneous maximum frontwidth.

35* Write the maximum frontwidth encountered in the problem.

36* If the maximum frontwidth is greater than the maximum permissible value allowed by subroutine FRONT set the error indicator for error type 21 equal to 1.

37* Enter loop over each restrained nodal point.

38* If the restrained nodal point number is less than or equal to zero or greater than the specified total number of nodes in the structure, increase the error counter for error type 22 by 1.

39* Set the counter, KOUNT, signifying a zero constraint code for the nodal point, equal to zero.

40* If the restraint code is greater than zero, set KOUNT = 1.

41* If the nodal restraint code is zero, increment the error counter for error type 23 by 1.

42* Enter loop over the restrained nodes up to, but not including, the node being currently checked.

43* Each time the number of the restrained node being currently considered is equal to that of any of the preceding nodes, increment the error counter for error type 24 by 1.

44* Initialise the indicator which will signal whether or not any errors have been detected in this subroutine.

45* Loop over each error type checked for in this subroutine.

46* If the error counter for a particular error type is zero, go to the end of the loop and process the next error type.

47* Otherwise, set the error indicator, KEROR, equal to 1 and output the error number together with the associated number. The associated number indicates how many times the particular error type has been detected.

48* If any errors have been detected branch to call subroutine ECHO.

49* Otherwise revert all the element nodal connection numbers to positive values and return to subroutine INPUT.

50* If errors have been detected call subroutine ECHO to list the remainder of the input data.

9.5 List of error diagnostics

As seen in the preceding sections any errors which are detected are signalled by the printing of an error number and an associated number, which indicates how many times the particular error type has been diagnosed. The interpretation of each error number is as indicated below

Diagnosed by Subroutine CHECK1

Error Label	Interpretation
1	The specified total number of node points, NPOIN, in the structure is less than or equal to zero.
2	The possible maximum total number of node points in the structure is less than the specified total, NPOIN.
3	The number of restrained nodal points is less than 1 or greater than NPOIN (for beam problems at least 1 point must be restrained to eliminate rigidy body motions)
4	The number of load cases is less than or equal to zero.
5	The problem type parameter, NTYPE, is not specified as either 0, 1 or 2.
6	The number of nodes/element is less than 3 (beam) or greater than 8 (plane, plate elements).
7	The number of degrees of freedom per node is not equal to 2 (beam, plane) or 3 (plate problems).
8	The total number of different materials is less than or equal to zero or greater than the total number of elements in the structure.
9	The number of properties per material is less than 3 (beam), or greater than 5 (plane problems).
10	The number of Gaussian integration points in each direction is not equal to either 2 or 3.
11	The number of coordinate dimensions is not equal to either 1 (beam) or 2 (plane, plate problems).
12	The size of the stress matrix is less than 2 (beam) or greater than 5 (plate problems).

Diagnosed by Subroutine CHECK2

Error Label	Interpretation
13	A total of x identical nodal coordinates have been detected, i.e. x nodal points have coordinates which are identical to those of one or more of the remaining nodes.
14	A total of x element material identification numbers are less than or equal to zero or greater than the total number of elements in the structure.
15	A total of x nodal connection numbers have a zero value.
16	A total of x nodal connection numbers are negative or greater than the specified maximum value, NPOIN.
17	A total of x repetitions of node numbers within individual elements have been detected.
18	A total of x nodes exist in the list of nodal points which do not appear anywhere in the list of element nodal connection numbers.
19	Non-zero coordinates have been specified for a total of x nodes which do not appear in the list of element nodal connection numbers.
20	A total of x node numbers which do not appear in the element nodal connections list have been specified as restrained nodal points.
21	The largest frontwidth encountered in the problem has exceeded the maximum value specified in subroutine FRONT of the program.
22	A total of x restrained nodal points have numbers less than or equal to zero or greater than the specified maximum value, NPOIN.
23	A total of x restrained nodal points at which the fixity code is less than or equal to zero have been detected.
24	A total of x repetitions in the list of restrained nodal points have been detected.

9.6 Illustrative examples

In this section a numerical example containing several deliberate data errors is presented. The problem chosen is a thick cylinder under internal pressure which is solved later as an illustrative example in Chapter 10. The geometry, loading and material properties are as shown in.Fig. 10.4 where the finite element mesh of 9 elements is also indicated.

Firstly, deliberate errors are introduced into the problem control parameter card to illustrate the error diagnostics generated by subroutine CHECK1. Other errors are introduced into the remainder of the data. The correct data sequence for this problem can be found in Section 10.4.2. With the incorrect data the lineprinter output obtained is that shown on pp. 220–222; the incorrect data values being indicated by an asterisk.

Since the program fails after checking the problem control parameters, these must be corrected before the remainder of the data can be checked by subroutine CHECK2. When this has been done the lineprinter output obtained is that shown on pp. 223–225.

TOTAL NO. OF PROBLEMS =   1

PROBLEM NO.   1

PLANE STRAIN THICK CYLINDER - WITH INTRODUCED ERRORS

NPOIN =   0*   NELEM =   9   NVFIX = 14   NCASE =   0*   NTYPE =   3*   NNODE = -1*   NDOFN =   0*
NMATS = 10*   NPROP =   2* NGAUS =   1*   NDIME =   3*   NSTRE =   6*   NEVAB = -0*

*** DIAGNOSIS BY CHECK1, ERROR 1

*** DIAGNOSIS BY CHECK1, ERROR 2

*** DIAGNOSIS BY CHECK1, ERROR 3

*** DIAGNOSIS BY CHECK1, ERROR 4

*** DIAGNOSIS BY CHECK1, ERROR 5

*** DIAGNOSIS BY CHECK1, ERROR 6

*** DIAGNOSIS BY CHECK1, ERROR 7

*** DIAGNOSIS BY CHECK1, ERROR 8

```
*** DIAGNOSIS BY CHECK1, ERROR  9

*** DIAGNOSIS BY CHECK1, ERROR 10

*** DIAGNOSIS BY CHECK1, ERROR 11

*** DIAGNOSIS BY CHECK1, ERROR 12

NOW FOLLOWS A LISTING OF POST-DISASTER DATA CARDS

 1    1    1    2    3    9   14   13   12    9
 2    1    3    4    7   10   16   15   14    9
 3    1    5    6   11   18    0   16   10
 4    0   12   13   14   20   25   24   23   49
 5    1   64   15   16   21   27   26   25   20
 6    1   16   17   18   22   29   28   27   21
 7   10   23   24   25   31   31   45   34   30
 8    1   45   26   27   32   38   37   36   31
 9         27   28   29   33   40   39   38    0

 1    5.0        0.0
 3    8.3333     0.0
 5   13.0        0.0
 7   20.0        0.0
 8    4.8296     1.2941
 9    4.8296     1.2941
10   12.5570     3.3646
11   19.3185     5.1764
11    4.3301     2.5
14    7.2169     4.1667
16    5.0        0.0
18   17.3205    10.0
19    3.5355     3.5355
20    5.8926     5.8926
```

```
21    9.1924     9.1924
22   14.1421    14.1421
23    2.5        4.3301
25    4.1667     7.2169
27    6.5       11.2583
29   10.0       17.3205
30    1.2941     4.8296
31    2.1568     8.0494
32    3.3646    12.5570
23    5.1764    19.3185
34    0.0        5.0
16    0.0        8.3333
38    0.0       13.0
40    0.0       20.0
 1   01
 2   01
 3   01
 4   01
 8   01
 6   01
 3   01
34   10
35   10
36   00
37   10
38   10
49   10
40   00
 1 1000.0    0.3      0.001
 0    1       0
 3    0
 1   12    8    1       1.0     0.0     0.0     0.0
10.0  10.0  10.0        0.0     0.0     0.0
 4   23   19   12
10.0  10.0  10.0        0.0     0.0     0.0
 7   34   30   23
10.0  10.0  10.0        0.0     0.0     0.0

INTERNAL PRESSURE LOADING
```

TOTAL NO. OF PROBLEMS =   1

PROBLEM NO.   1   PLANE STRAIN THICK CYLINDER - WITH INTRODUCED ERRORS

NPOIN = 40   NELEM =   9   NVFIX = 14   NCASE =   1   NTYPE =   2   NNODE =   8   NDOFN =   2

NMATS =   1   NPROP =   5   NGAUS =   3   NDIME =   2   NSTRE =   3   NEVAB = 16

ELEMENT	PROPERTY	NODE NUMBERS							
1	1	1	2	3*	9	14	13	12	9*
2	1	3	4	5	10	16	15	14	9
3	1	5	6	7	11	18	0*	16	10
4	0*	12	13	14	20	25	24	23	49*
5	1	64*	15	16	21	27	26	25	20
6	1	16	17	18	22	29	28	27	21
7	10*	23	24	25	31	31*	45*	34	30
8	1	45*	26	27	32	38	37	36	31
9	1	27	28	29	33	40	39	38	0*

NODAL POINT COORDINATES

NODE	X	Y
1	5.000	0.000
2	6.667	0.000
3	8.333	0.000
4	8.333*	0.000
5	13.000	0.000
6	16.500	0.000
7	20.000	0.000
8	4.830	1.294
9	4.830*	1.294
10	12.557	3.365
11	4.330*	2.500
12	0.000*	2.500*
13	3.608*	2.083*
14	7.217	4.167
15	3.608*	6.250*
16	0.000*	8.333*
17	8.660*	9.167*
18	17.321	10.000
19	3.536	3.536
20	5.893	5.893
21	9.192	9.192
22	14.142	14.142

```
23   5.176*   19.319*
24   4.672*   13.268**
25   4.167     7.217
26   5.333     9.238
27   6.500    11.258
28   8.250    14.289
29  10.000    17.321
30   1.294     4.830
31   2.157     8.049
32   3.365    12.557
33   5.000*   18.660*
34   0.000     5.000
35   0.000*    0.000**
36   0.000     0.000***
37   0.000     6.500*
38   0.000    13.000
39   0.000    16.500
40   0.000    20.000
```

RESTRAINED NODES
NODE CODE   FIXED VALUES

```
 1   01   -0.00000   -0.000000
 2   01   -0.000000  -0.000000
 3   01   -0.000000  -0.000000
 4   01   -0.000000  -0.000000
 8*  01   -0.000000  -0.000000
 6   01   -0.000000  -0.000000
 3*  01   -0.000000  -0.000000
34   10   -0.000000  -0.000000
35   00*  -0.000000  -0.000000
36   00*  -0.000000  -0.000000
37   10   -0.000000  -0.000000
38   10   -0.000000  -0.000000
49*  10   -0.000000  -0.000000
40   00*  -0.000000  -0.000000
```

MATERIAL PROPERTIES
NUMBER   PROPERTIES-
 1   .100000E+04   .300000E+00   .100000E+01   0.   .100000E-02

CHECK WHY NODE   8 NEVER APPEARS

CHECK WHY NODE 19 NEVER APPEARS

CHECK WHY NODE 35 NEVER APPEARS

```
MAXIMUM FRONTWIDTH ENCOUNTERED =   24

*** DIAGNOSIS BY CHECK2, ERROR 13     ASSOCIATED NUMBER     5

*** DIAGNOSIS BY CHECK2, ERROR 14     ASSOCIATED NUMBER     2

*** DIAGNOSIS BY CHECK2, ERROR 15     ASSOCIATED NUMBER     2

*** DIAGNOSIS BY CHECK2, ERROR 16     ASSOCIATED NUMBER     4

*** DIAGNOSIS BY CHECK2, ERROR 17     ASSOCIATED NUMBER     3

*** DIAGNOSIS BY CHECK2, ERROR 18     ASSOCIATED NUMBER     3

*** DIAGNOSIS BY CHECK2, ERROR 19     ASSOCIATED NUMBER     2

*** DIAGNOSIS BY CHECK2, ERROR 20     ASSOCIATED NUMBER     2

*** DIAGNOSIS BY CHECK2, ERROR 22     ASSOCIATED NUMBER     1

*** DIAGNOSIS BY CHECK2, ERROR 23     ASSOCIATED NUMBER     2

*** DIAGNOSIS BY CHECK2, ERROR 24     ASSOCIATED NUMBER     1

NOW FOLLOWS A LISTING OF POST-DISASTER DATA CARDS

     INTERNAL PRESSURE LOADING
      0     1     0     1
      3    12     8     1
     10.0  10.0  12          10.0    0.0    0.0
      4    23    19    12
     10.0  10.0  10.0        10.0    0.0    0.0
      7    34    30    23
     10.0  10.0  10.0        10.0    0.0    0.0
```

Again the reader can easily relate the diagnostic messages to the incorrect data values. The program fails, as before, as soon as the remaining input cards (the loading data) have been echoed by the lineprinter.

It is worth re-emphasising that the error diagnostic subroutines presented will only detect a major proportion of the errors made in preparing input data. As far as the structural geometry is concerned a graphical plot of the finite element mesh gives a far better data check. However, diagnostic subroutines have the advantage that the data is automatically scrutinised and that peripheral equipment and software are not required as is the case for graphical display. Furthermore the subroutines presented also check other items of data, such as the control parameters, the frontwidth of the problem and the boundary conditions.

It is possible to write more sophisticated data checking subroutines in which the diagnostics relating to array dimensioning are an important feature. Some alternative schemes for scrutinising the input data to ensure that the maximum array dimensions are not exceeded are discussed in Chapter 11, Section 11.13.

References

1. Irons, B. M., "Error Diagnostics and Dynamic Dimensioning". Dept of Civil Engineering, University of Calgary, Internal Report. 1975.

10

Program Construction and Numerical Examples

10.1 Introduction

We have now developed all of the subroutines required for the three programs with one exception in each case: the master or controlling subroutine. In this chapter we describe the master subroutine and explain how the other various subroutines which have been described earlier are assembled to produce the three programs. The common block organisation which had been omitted previously to avoid confusion is also described in detail for each program. In the last part of the chapter some test examples are presented illustrating the use of each program. These examples should be useful for those readers who wish to implement the programs described in the text or for those more ambitious readers who prefer to write their own versions of these programs

10.2 Common block organisation

Three common blocks are used to store information in each of the finite element programs. The first common block, which is called CONTRO, stores control parameters which are defined in subroutine INPUT and which are used throughout the program. Common block CONTRO has the following form:

    COMMON/CONTRO/NPOIN,NELEM,NNODE,NDOFN,NDIME,
        NSTRE,NTYPE,NGAUS,NPROP,NMATS,NVFIX,NEVAB,
        ICASE,NCASE,ITEMP,IPROB,NPROB

This common block is used in the above form in *all* subroutines in each of the three programs.

The second common block which is called LGDATA stores a set of arrays which have different dimensions in each of the three programs. First we will examine the general form of LGDATA which has the following general form:

COMMON/LGDATA/COORD(MPOIN, NDIME),PROPS(MMATS, NPROP),PRESC(MVFIX,NDOFN),ASDIS(NTOTV),ELOAD (MELEM,NEVAB),[STRIN(NSTRE,NTOTG)],NOFIX(MVFIX), IFPRE(MVFIX,NDOFN),LNODS(MELEM,NNODE),MATNO (MELEM)

where the term in square brackets only occurs in the plane stress/strain application.

For the three programs the dimensions have the following values.

	Beam	Plane Stress/Strain	Plate Bending
MPOIN	50	80	80
NDIME	1	2	2
MMATS	10	10	10
NPROP	3	5	4
MVFIX	50	40	40
NDOFN	2	2	3
NTOTV	100	160	240
(=MPOIN*NDOFN)			
MELEM	25	25	25
NEVAB	6	16	24
NSTRE	2	3	5
NTOTG = (MELEM*NGASP)	—	225	—
NGASP	—	9	—

These dimension values can be varied for each program—however if they are changed care should be taken to ensure that all of the implications of such changes are considered.

i.e. Common block LGDATA should be changed where appropriate and the DIMENSION statement in subroutine FRONT should be modified accordingly (see Chapter 8) as well as the dimension of NDFRO in subroutine CHECK2 (see Chapter 9). The third common block which is called WORK stores a set of arrays which specifically relates to the application under consideration and has the following general form:

COMMON/WORK/ELCOD(NDIME,NNODE),SHAPE(NNODE),

DERIV(NDIME,NNODE),DMATX(NSTRE,NSTRE),CARTD
(NDIME,NNODE),DBMAT(NSTRE,NEVAB),BMATX(NSTRE,
NEVAB),SMATX(NSTRE,NEVAB,NGASP),POSGP(NGAUS)
WEIGP(NGAUS),GPCOD(NDIME,NGASP),NEROR(24).

For the three programs the dimensions have the following values

	Beam	Plane Stress/Strain	Plate Bending
NDIME	1	2	2
NNODE	3	8	8
NSTRE	2	3	5
NEVAB	6	16	24
NGASP	2	9	4
NGAUS	2	3	2

10.3 The assembled programs

The general form of the assembled programs has been given in Fig. 1.7 and it can be seen that one routine is used to control the whole finite element analysis. An annotated listing of this master routine together with appropriate common blocks and associated primary and auxiliary subroutine headings is now presented for each program.

First for the beam application we have:

Annotated FORTRAN listing

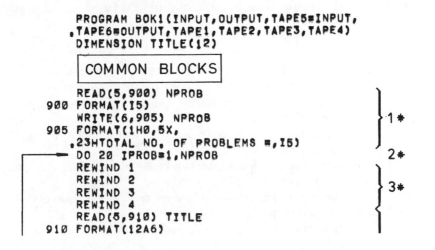

```
      PROGRAM BOK1(INPUT,OUTPUT,TAPE5=INPUT,
     .TAPE6=OUTPUT,TAPE1,TAPE2,TAPE3,TAPE4)
      DIMENSION TITLE(12)

      | COMMON BLOCKS |

      READ(5,900) NPROB
  900 FORMAT(I5)
      WRITE(6,905) NPROB                          } 1*
  905 FORMAT(1H0,5X,
     .23HTOTAL NO. OF PROBLEMS =,I5)
      DO 20 IPROB=1,NPROB                           2*
      REWIND 1
      REWIND 2
      REWIND 3                                     } 3*
      REWIND 4
      READ(5,910) TITLE
  910 FORMAT(12A6)
```

```
        WRITE(6,915) IPROB,TITLE                    ⎫ 4*
    915 FORMAT(/////,6X,12HPROBLEM NO. ,I3,10X,     ⎬
       .12A6)                                       ⎭
      C
      C*** CALL THE SUBROUTINE WHICH READS MOST OF
      C    THE PROBLEM DATA
      C
            CALL INPUT                                5*
      C
      C*** NEXT CREATE THE ELEMENT STIFFNESS FILE.
      C
            CALL STIFB                                6*
   ┌──     DO 10 ICASE=1,NCASE                        7*
   │  C
   │  C*** COMPUTE LOADS, AFTER READING
   │  C    THE RELEVANT EXTRA DATA
   │  C
   │        CALL LOADB                                8*
   │  C
   │  C*** MERGE AND SOLVE THE RESULTING EQUATIONS
   │  C    BY THE FRONTAL SOLVER
   │  C
   │        CALL FRONT                                9*
   │  C
   │  C*** COMPUTE THE STRESSES IN ALL THE ELEMENTS.
   │  C
   │        CALL STREB                               10*
   └── 10 CONTINUE                                   11*
   └── 20 CONTINUE                                   12*
          STOP
          END
```

1* Read and write the total number of problems to be solved in one computer run.
2* Loop over each problem.
3* Rewind files ready for writing.
4* Read and write title associated with the current problem.
5* Call the subroutine which reads most of the input data.
6* Call the subroutine which creates the element stiffness matrix, etc.
7* Loop over each load case.
8* Call the subroutine which computes the loads after reading the relevant load data.
9* Call the subroutine which assembles and solves the resulting equations by the frontal solver.
10* Call the subroutine which computes the stresses in all elements.
11* End of load case loop.
12* End of problem loop.

	Name	*Section in which it is described and listed*
Subroutine	*Name*	

SUBROUTINE	INPUT	3.7
	NODEXY	3.9
	GAUSSQ	3.10
	STIFB	4.10.1
	MODB	4.10.5
	SFR1	4.10.2
	JACOB1	4.10.3
	BMATB	4.10.4
	DBE	4.10.6
	LOADB	4.10.7
	FRONT	8
	STREB	4.10.8
	CHECK1	9.2
	ECHO	9.3
	CHECK2	9.4

For the plane stress/strain program we have

Annotated FORTRAN listing

```
      PROGRAM BOK1(INPUT,OUTPUT,TAPE5=INPUT,
     .TAPE6=OUTPUT,TAPE1,TAPE2,TAPE3,TAPE4)
      DIMENSION TITLE(12)
```

┌─────────────────────┐
│ COMMON BLOCKS │
└─────────────────────┘

```
      READ(5,900) NPROB
  900 FORMAT(I5)
      WRITE(6,905) NPROB
  905 FORMAT(1H0,5X,23HTOTAL NO. OF PROBLEMS =,I5)
      DO 20 IPROB=1,NPROB
      REWIND 1
      REWIND 2
      REWIND 3
      REWIND 4
      READ(5,910) TITLE
  910 FORMAT(12A6)
      WRITE(6,915) IPROB,TITLE
  915 FORMAT(/////,6X,12HPROBLEM NO. ,I3,10X,12A6)
C
C*** CALL THE SUBROUTINE WHICH READS MOST OF
C    THE PROBLEM DATA
C
      CALL INPUT
C
C*** NEXT CREATE THE ELEMENT STIFFNESS FILE.
C
```

```
         CALL STIFPS
      ┌──DO 10 ICASE=1,NCASE
    C
    C*** COMPUTE LOADS, AFTER READING THE RELEVANT
    C    EXTRA DATA
    C
         CALL LOADPS
    C
    C*** MERGE AND SOLVE THE RESULTING EQUATIONS
    C    BY THE FRONTAL SOLVER
    C
         CALL FRONT
    C
    C*** COMPUTE THE STRESSES IN ALL THE ELEMENTS.
    C
         CALL STREPS
    ├── 10 CONTINUE
    └── 20 CONTINUE
         STOP
         END
```

Subroutine	Name	Section in which it is described and listed
SUBROUTINE	INPUT	3.7
	NODEXY	3.9
	GAUSSQ	3.10
	STIFPS	5.13
	MODPS	5.9
	SFR2	5.3
	JACOB2	5.5
	BMATPS	5.7
	DBE	4.10.6
	LOADPS	7.5
	FRONT	8
	STREPS	5.15
	CHECK1	9.2
	ECHO	9.3
	CHECK2	9.4

For the plate bending program we have

Annotated FORTRAN listing

```
      PROGRAM BOK1(INPUT,OUTPUT,TAPE5=INPUT,
     .TAPE6=OUTPUT,TAPE1,TAPE2,TAPE3,TAPE4)
      DIMENSION TITLE(12)
```

┌─────────────────────┐
│ COMMON BLOCKS   │
└─────────────────────┘

```
      READ(5,900) NPROB
  900 FORMAT(I5)
      WRITE(6,905) NPROB
  905 FORMAT(1H0,5X,
     .23HTOTAL NO. OF PROBLEMS =,I5)
      DO 20 IPROB=1,NPROB
      REWIND 1
      REWIND 2
      REWIND 3
      REWIND 4
      READ(5,910) TITLE
  910 FORMAT(12A6)
      WRITE(6,915) IPROB,TITLE
  915 FORMAT(/////,6X,12HPROBLEM NO. ,I3,10X,
     .12A6)
C
C*** CALL THE SUBROUTINE WHICH READS MOST OF
C    THE PROBLEM DATA
C
      CALL INPUT
C
C*** NEXT CREATE THE ELEMENT STIFFNESS FILE.
C
      CALL STIFPB
      DO 10 ICASE=1,NCASE
C
C*** COMPUTE LOADS, AFTER READING THE RELEVANT
C    EXTRA DATA
C
      CALL LOADPB
C
C*** MERGE AND SOLVE THE RESULTING EQUATIONS
C    BY THE FRONTAL SOLVER
C
      CALL FRONT
C
C*** COMPUTE THE STRESSES IN ALL THE ELEMENTS.
C
      CALL STREPB
   10 CONTINUE
   20 CONTINUE
      STOP
      END
```

Subroutine	Name	*Section in which it is described and listed*
SUBROUTINE	INPUT	3.7
	NODEXY	3.9
	GAUSSQ	3.10
	STIFPB	6.9
	MODPB	6.6
	SFR2	5.3
	JACOB2	5.5
	BMATPB	6.4
	DBE	4.10.6
	LOADPB	7.6
	FRONT	8
	STREPB	6.11
	CHECK1	9.2
	ECHO	9.3
	CHECK2	9.4

10.4 Numerical examples

The numerical examples presented here will be essentially simple, in order that comparisons with available analytical solutions can be made. The consideration of more complicated situations is left to the reader who it is expected will use the programs to solve his own practical problems.

In order to enable the reader to check initially the developed programs, ouput listings for a selection of the problems solved are presented. In particular output for one beam problem, one plane stress/strain analysis and one plate bending solution are included.

10.4.1 Beam analysis

The beam example is the simply supported beam shown in Fig. 10.1 where the beam dimensions and the material properties assumed are also indicated. The beam is subjected to a uniformly distributed load, $q = 10$. This problem is solved using two parabolic isoparametric elements as illustrated in Fig. 10.1. The results of the analysis are given below:

$$\text{Central deflection} = 0.0130$$
$$\text{Bending moment at Gauss point B} = 0.119$$
$$\text{Shear Force at Gauss point B} = -0.106$$

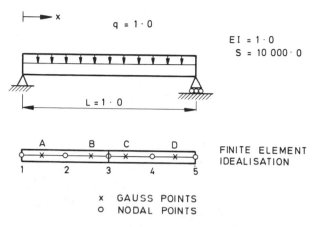

FIG. 10.1.   Test example for beam element.

These results are in excellent agreement with those predicted by simple beam theory.

10.4.2 *Plane stress/plane strain applications*

The plane stress application considered is the cantilever beam shown in Fig. 10.2 where the beam dimensions and the material properties assumed are also indicated. The beam is subjected to two forms of loading; namely a concentrated load applied at the free end and a self-weight gravitational loading. The problem is solved using 10 parabolic isoparametric elements as illustrated in Fig. 10.2. The axial stress distribution at different sections along the beam is compared with the theoretical values, based on simple beam bending theory, in Fig. 10.3 for the concentrated loading case. The end deflections caused by both loadings are also compared with the theoretical values, the comparison being included in Fig. 10.3

   This simple problem illustrates the efficiency of the parabolic isoparametric element, since from Fig. 10.3 it is seen that excellent axial stress prediction is obtained with only one element employed in the transverse direction; simple constant stress elements require a much finer subdivision for comparable accuracy. (This accuracy is to be expected since the true transverse stress distribution is known to be linear.) Furthermore from Fig. 10.3, it is seen that the numerical values for the end deflection are within 1 % of the theoretical value even for the relatively coarse subdivision employed.

   The other two-dimensional solid example considered is the case of a thick circular cylinder conforming to plane strain conditions. Fig. 10.4 shows the geometrical dimensions and the material properties assumed. The element

OUTPUT FOR BEAM TEST PROBLEM

TOTAL NO. OF PROBLEMS = 1

PROBLEM NO. 1   ** BEAM TEST EXAMPLE **

NPOIN = 5   NELEM = 2   NVFIX = 2   NCASE = 1   NTYPE = 0   NNODE = 3   NDOFN = 2

NMATS = 1   NPROP = 3   NGALS = 2   NDIME = 2   NSTRE = 2   NEVAB = 6

```
ELEMENT   PROPERTY    NODE NUMBERS
   1         1          1   2   3
   2         1          3   4   5

NODAL POINT COORDINATES
NODE      X        Y
  1     0.000
  2      .250
  3      .500
  4      .750
  5     1.000

RESTRAINED NODES
NODE CODE     FIXED VALUES
  1   10   -0.000000  -0.000000
  5   10   -0.000000  -0.000000

MATERIAL PROPERTIES
NUMBER       PROPERTIES
   1      .100000E+01   .100000E+05   .100000E+01
```

```
MAXIMUM FRONTWIDTH ENCOUNTERED =   6

   TOTAL NODAL FORCES FOR EACH ELEMENT
  1   .8333E-01  0.  .8333E-01  0.  .3333E+00
  2   0.  .8333E-01  0.  .8333E-01  .3333E+00
      0.  .8333E-01  0.

  DISPLACEMENTS

NODE     DISP.        ROTATION
  1  =0.             .416667E-01
  2   .912396E-02    .286458E-01
  3   .130333E-01    .340995E-14
  4   .912396E-02  -.286458E-01
  5  =0.            -.416667E-01

  REACTIONS

NODE     FORCE         MOMENT
  1  =-.500000E+00  =0.
  5  =-.500000E+00  =0.

  STRESSES

G.P.  X-COORD  X-MOMENT  XZ-S.FORCE

  ELEMENT NO.=   1
  1   .1057  =-.47249E-01  =-.39434E+00
  2   .3943  =-.11942E+00  =-.10566E+00

  ELEMENT NO.=   2
  1   .6057  =-.11942E+00   .10566E+00
  2   .8943  =-.47249E-01   .39434E+00
```

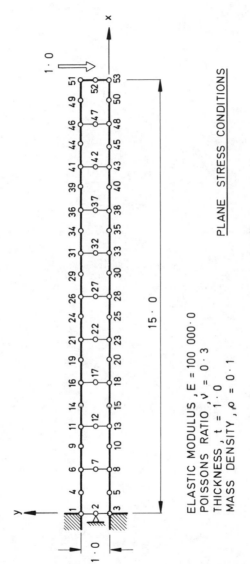

ELASTIC MODULUS , E = 100 000·0
POISSONS RATIO , ν = 0·3
THICKNESS , t = 1·0
MASS DENSITY , ρ = 0·1

PLANE STRESS CONDITIONS

Fɪɢ. 10.2.   Cantilever beam test example—plane stress.

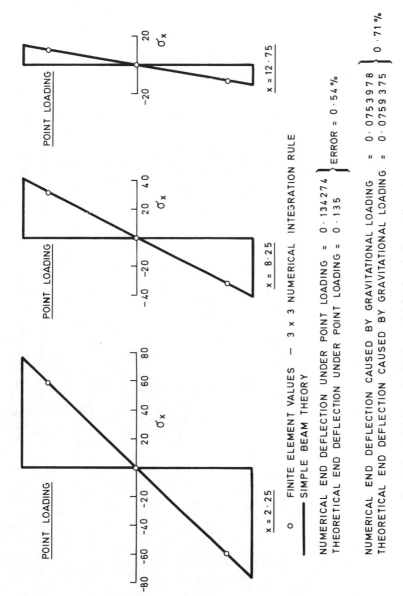

FIG. 10.3.   Axial stress distribution and deflections in the cantilever beam.

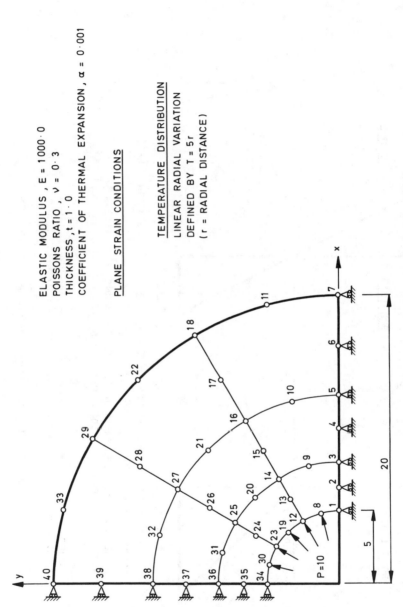

ELASTIC MODULUS , E = 1000·0
POISSONS RATIO , ν = 0·3
THICKNESS , t = 1·0
COEFFICIENT OF THERMAL EXPANSION , α = 0·001

PLANE STRAIN CONDITIONS

TEMPERATURE DISTRIBUTION
LINEAR RADIAL VARIATION
DEFINED BY T = 5 r
(r = RADIAL DISTANCE)

FIG. 10.4. Thick circular cylinder test example—plane strain.

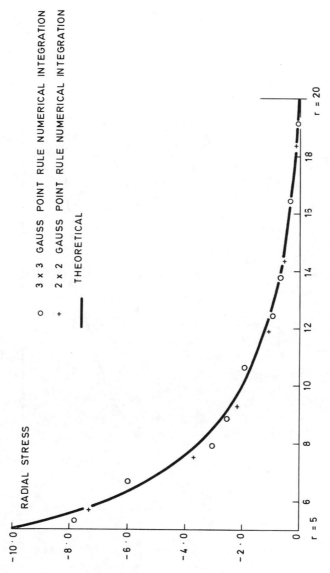

FIG. 10.5(a). Radial stress distribution due to internal pressure loading.

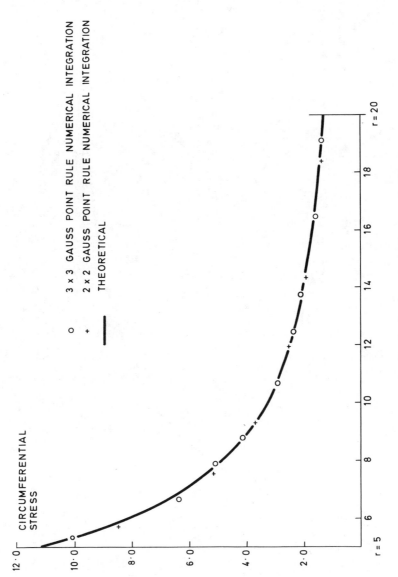

FIG. 10.5(b).   Hoop stress distribution due to internal pressure loading.

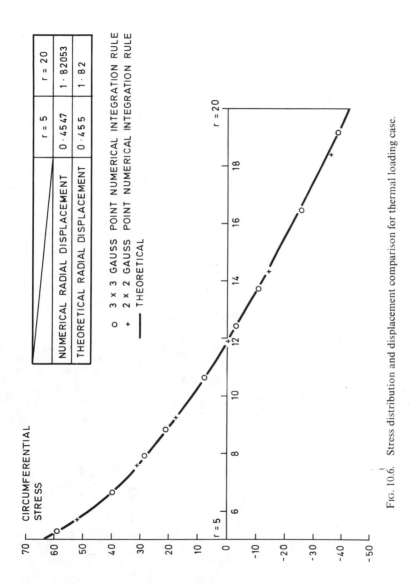

FIG. 10.6. Stress distribution and displacement comparison for thermal loading case.

OUTPUT FOR PLANE STRAIN THICK CYLINDER TEST PROBLEM

TOTAL NO. OF PROBLEMS = 1

PROBLEM NO. 1   THICK CYLINDER UNDER PLANE STRAIN CONDITIONS

NPOIN = 40   NELEM = 9   NVFIX = 14   NCASE = 2   NTYPE = 2   NNODE = 8   NDOFN = 2

NMATS = 1   NPROP = 5   NGAUS = 2   NDIME = 2   NSTRE = 3   NEVAB = 16

ELEMENT	PROPERTY	NODE NUMBERS							
1	1	1	2	3	9	14	13	12	8
2	1	3	4	5	10	16	15	14	9
3	1	5	6	7	11	18	17	16	10
4	1	12	13	14	20	25	24	23	19
5	1	14	15	16	21	27	26	25	20
6	1	16	17	18	22	29	28	27	21
7	1	23	24	25	31	36	35	34	30
8	1	25	26	27	32	38	37	36	31
9	1	27	28	29	33	40	39	38	32

NODAL POINT COORDINATES

NODE	X	Y
1	5.000	0.000
2	6.667	0.000
3	8.333	0.000
4	10.667	0.000
5	13.000	0.000
6	16.500	0.000
7	20.000	0.000
8	4.830	1.294
9	8.049	2.157
10	12.557	3.365

11	19.319	5.176
12	4.330	2.500
13	5.774	3.333
14	7.217	4.167
15	9.238	5.333
16	11.258	6.500
17	14.289	8.250
18	17.321	10.000
19	3.536	3.536
20	5.893	5.893
21	9.192	9.192
22	14.142	14.142
23	2.500	4.330
24	3.333	5.774
25	4.167	7.217
26	5.333	9.238
27	6.500	11.258
28	8.250	14.289
29	10.000	17.321
30	1.294	4.830
31	2.157	8.049
32	3.365	12.557
33	5.176	19.319
34	0.000	5.000
35	0.000	6.667
36	0.000	8.333
37	0.000	10.667
38	0.000	13.000
39	0.000	16.500
40	0.000	20.000

RESTRAINED NODES

NODE	CODE	FIXED	VALUES
1	01	0.000000	0.000000
2	01	0.000000	0.000000
3	01	0.000000	0.000000
4	01	0.000000	0.000000
5	01	0.000000	0.000000
6	01	0.000000	0.000000
7	01	0.000000	0.000000
34	10	0.000000	0.000000

```
35    10    0.000000    0.000000
36    10    0.000000    0.000000
37    10    0.000000    0.000000
38    10    0.000000    0.000000
39    10    0.000000    0.000000
40    10    0.000000    0.000000

MATERIAL PROPERTIES
NUMBER      PROPERTIES
   1        .100000E+04    .300000E+00    .100000E+01    0.    .100000E-02

MAXIMUM FRONTWIDTH ENCOUNTERED =    24

   INTERNAL PRESSURE LOADING
0    0    1    0

NO. OF LOADED EDGES =    3

LIST OF LOADED EDGES AND APPLIED LOADS
   1      12       8       1
10.000   10.000   0.000    0.000    0.000
   4      23      19      12
10.000   10.000   0.000    0.000    0.000
   7      34      30      23
10.000   10.000   0.000    0.000    0.000

                                                              LOAD CASE =    1

TOTAL NODAL FORCES FOR EACH ELEMENT
1    .4461E+01    .1950E-01    0.    0.    0.    0.    .3873E+01    0.
2    0.    0.    0.    0.    0.    0.    0.    0.
3    0.    0.    0.    0.    0.    0.    0.    0.
4    .3853E+01    0.    .2247E+01    0.    0.    .2214E+01    .1667E+02    0.    .4466E+01
5    0.    0.    0.    0.    0.    0.    0.    0.
6    0.    0.    0.    0.    0.    0.    0.    .3853E+01    .1220E+02    .1220E+02
```

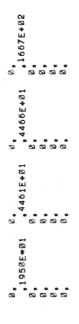

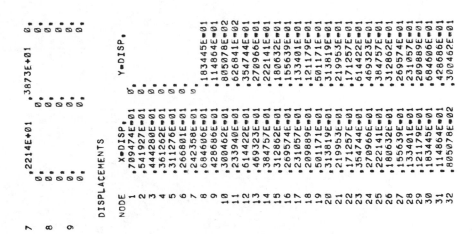

```
33    .626841E+02    .233940E-01
34  0.               .709474E-01
35  0.               .541927E-01
36  0.               .444280E-01
37  0.               .361262E-01
38  0.               .311276E-01
39  0.               .266801E-01
40  0.               .242358E-01
```

REACTIONS

```
NODE    X-FORCE          Y-FORCE
 1    0.              -.612866E+01
 2    0.              -.150709E+02
 3    0.              -.580855E+01
 4    0.              -.946415E+01
 5    0.              -.426964E+01
 6    0.              -.773523E+01
 7    0.              -.152285E+01
34  -.612866E+01     0.
35  -.150709E+02     0.
36  -.580855E+01     0.
37  -.946415E+01     0.
38  -.426964E+01     0.
39  -.773523E+01     0.
40  -.152285E+01     0.
```

STRESSES

ELEMENT NO.= 1

G.P.	X-COORD.	Y-COORD.	X-STRESS	Y-STRESS	XY-STRESS	Z-STRESS	MAX P.S.	MIN P.S.	ANGLE
1	5.3653	.3219	.78136E+01	.10158E+02	-.10936E+01	.70340E+00	.10225E+02	-.78799E+02	3.4694
2	5.1925	1.3913	.65711E+01	.90628E+01	-.45132E+01	.74752E+00	.10272E+02	-.77804E+02	15.0002
3	4.8074	2.4039	-.42675E+01	.66121E+01	-.72352E+01	.70336E+00	.10224E+02	-.78798E+02	26.5311
4	6.6538	.3992	.59675E+01	.63263E+01	.72432E+00	.10765E+00	.63689E+01	-.60100E+01	3.3602
5	6.4395	1.7255	.51816E+01	.55287E+01	-.30918E+01	.10416E+00	.63572E+01	-.60100E+01	15.0001
6	5.9619	2.9812	-.35214E+01	.38800E+01	-.49612E+01	.10757E+00	.63686E+01	-.60101E+01	26.6397
7	7.9423	.4765	-.30252E+01	.50154E+01	-.49074E+00	.59708E+00	.50452E+01	-.30550E+01	3.4797
8	7.6865	2.0596	.25935E+01	.44636E+01	-.20372E+01	.56103E+00	.50095E+01	-.31394E+01	14.9999
9	7.1165	3.5585	.14401E+01	.34303E+01	-.32362E+01	.59704E+00	.50452E+01	-.30550E+01	26.5197

ELEMENT NO. = 2

1	8.8421	.25146E+01	.41267E+01	.40625E+00	.48363E+00	.41515E+01	−.25394E+01	3.4878
2	8.5574	−.20556E+01	−.37234E+01	−.16682E+00	.50023E+00	.41704E+01	−.25026E+01	14.9998
3	7.9228	.12062E+01	.28183E+01	.26726E+01	.48364E+00	.41515E+01	−.25394E+01	26.5117
4	10.6460	.18616E+01	.29026E+01	.28375E+00	.31229E+00	.29194E+01	−.18784E+01	3.3965
5	10.3032	.15542E+01	−.25970E+01	−.11983E+01	.31282E+00	.29181E+01	−.18755E+01	14.9998
6	9.5391	−.91631E+00	.19573E+01	−.19210E+01	.31230E+00	.29194E+01	−.18784E+01	26.6031
7	12.4499	.88672E+00	.24183E+01	.20147E+00	.45949E+00	.24306E+01	−.89896E+00	3.4756
8	12.0490	.70092E+00	.21956E+01	.83613E+01	.44840E+00	.24196E+01	−.92496E+00	14.9998
9	11.1554	.23497E+00	−.17666E+01	.13304E+01	.45948E+00	.24306E+01	−.89897E+00	26.5240

ELEMENT NO. = 3

1	13.7622	.65094E+00	.20928E+01	.16684E+00	.43255E+00	.21029E+01	−.66105E+00	3.4671
2	13.3190	−.46305E+00	−.19246E+01	−.68925E+00	.43846E+00	.21093E+01	−.64773E+00	14.9999
3	12.3313	−.10951E+01	−.15514E+01	−.11046E+01	.43255E+00	.21029E+01	−.66105E+00	26.5326
4	16.4681	.38446E+00	.16033E+01	.11831E+00	.36565E+00	.16103E+01	−.39148E+00	3.3944
5	15.9378	.25749E+00	.14751E+01	.50015E+00	.36528E+00	.16091E+01	−.39148E+00	15.0000
6	14.7558	.10013E−01	.12088E+01	−.85355E−01	.36565E+00	.16103E+01	−.39148E+00	26.6055
7	19.1739	.80909E−02	−.14036E+01	.35322E+00	.42350E+00	.14088E+01	−.28495E−01	3.4871
8	18.5565	−.85840E−01	.13094E+01	.56163E+00	.41858E+00	.14041E+01	−.88043E−02	15.0000
9	17.1803	−.28302E+00	−.11286E+01	.23302E+00	.42350E+00	.14088E+01	−.28421E−02	26.5132

ELEMENT NO. = 4

1	2.9614	.23736E+01	.47179E+01	.83287E+01	.70327E+00	.10224E+02	−.78799E+01	33.4696
2	3.8011	.12456E+01	.12456E+01	−.90201E+01	.74733E+00	.10272E+02	−.77806E+01	45.0000
3	2.9614	.47179E+01	−.23736E+01	−.83287E+01	.70327E+00	.10224E+02	−.78799E+01	33.4696
4	3.6726	.22667E+01	.26254E+01	.56855E+01	.10761E+00	.63687E+01	−.60100E+01	33.3605
5	4.7141	.17355E+01	−.17355E+01	−.61836E+01	.10413E+00	.63577E+01	−.60100E+01	45.0000
6	3.6726	.26254E+01	.22667E+01	.56855E+01	.10761E+00	.63687E+01	−.60100E+01	33.3605
7	6.6400	.50006E+00	−.25802E+01	−.37270E+01	.59006E+00	.50452E+01	−.50550E+01	33.4798
8	5.6270	.93503E+00	.93503E+00	.40744E+01	.56102E+00	.50094E+01	−.51394E+01	45.0000
9	4.3838	.25802E+01	.59006E+01	−.37270E+01	.59704E+00	.50452E+01	−.50550E+01	−33.4798

ELEMENT NO. = 5

1	4.8805	.50243E+00	.21144E+01	.30789E+01	.48361E+00	.41514E+01	−.25394E+01	33.4881
2	6.2645	.83387E+00	.83387E+00	.33355E+01	.50032E+00	.41704E+01	−.25026E+01	45.0000
3	7.3923	.21144E+01	.50243E+01	.30789E+01	.48361E+00	.41514E+01	−.25394E+01	33.4881
4	5.8762	.42482E+00	.14658E+01	.22048E+01	.31228E+00	.29181E+01	−.18784E+01	33.3966
5	7.5425	.52138E+00	.52138E+00	.23967E+01	.31283E+00	.29181E+01	−.18753E+01	45.0000
6	8.9004	.14658E+01	.42482E+00	.23967E+01	.31228E+00	.24306E+01	−.18784E+01	33.3966
7	6.8718	.11402E+01	.14176E+01	.15319E+01	.45947E+00	.24306E+01	−.89898E+00	33.4757
8	8.8205	.74735E+00	.74735E+00	.16723E+01	.44841E+00	.24196E+01	−.92495E+00	45.0000
9	10.4085	.14176E+01	.11402E+01	.15319E+01	.45947E+00	.24306E+01	−.89898E+00	−33.4757

ELEMENT NO. = 6

#							
1	11.5056	.17947E+00	.12624E+01	.12715E+01	.43255E+00	.21029E+01	.66105E+00
2	9.7502	.73078E+00	.73078E+00	.13785E+01	.43847E+00	.21093E+01	-.64773E+00
3	7.5962	.12624E+01	.17947E+01	.12715E+01	.43255E+00	.21029E+01	.66105E+00
4	13.7678	.21494E+00	.60881E+00	.91988E+00	.36565E+00	.16103E+01	-.39148E+00
5	11.6673	.60881E+00	.21494E+01	.10003E+01	.36528E+00	.16091E+01	.39151E+00
6	9.0897	.10039E+01	.10039E+01	.91988E+00	.36665E+00	.16103E+01	.39148E+00
7	16.0299	.43088E+00	.98082E+00	.64699E+00	.42351E+00	.14088E+01	.28517E-02
8	13.5843	.69764E+00	.70644E+00	.70644E+00	.41858E+00	.14041E+01	-.87988E-02
9	10.5832	.98082E+00	.43088E+01	.64699E+00	.42351E+00	.14088E+01	.28517E-02

#	
1	33.4671
2	45.0000
3	-33.4671
4	33.3944
5	45.0000
6	-33.3944
7	33.4872
8	45.0000
9	-33.4872

ELEMENT NO. = 7

#							
1	4.8074	.66121E+01	.42675E+01	.72352E+01	.70336E+00	.10224E+02	.78798E+01
2	5.1925	.90628E+01	.65711E+01	.45132E+01	.74752E+00	.10272E+02	-.77804E+01
3	5.3653	.10158E+02	.78136E+02	.10936E+01	.70340E+00	.10225E+02	.78799E+01
4	5.9619	.38800E+01	.35214E+01	.49618E+01	.10757E+01	.63686E+01	-.60101E+01
5	6.4395	.51816E+01	.30918E+01	.30918E+01	.10416E+01	.63578E+01	.60101E+01
6	6.6538	.63263E+01	.72432E+00	.72432E+00	.10765E+01	.63689E+01	.60100E+01
7	7.1165	.34303E+01	.14401E+01	.32362E+01	.59704E+00	.59452E+01	-.30555E+01
8	7.6865	.44636E+01	.25935E+01	.20372E+01	.56103E+00	.50095E+01	.31394E+01
9	7.9423	.50154E+01	.30252E+01	.49074E+01	.59708E+00	.59452E+01	.30550E+01

#	
1	-26.5311
2	15.0002
3	-3.4694
4	-26.6397
5	15.0001
6	-3.3602
7	-26.5197
8	-14.9999
9	-3.4797

ELEMENT NO. = 8

#							
1	3.9617	.28183E+01	.12062E+01	.26726E+01	.48364E+00	.41515E+01	.25394E+01
2	8.5574	.37234E+01	.20556E+01	.16682E+01	.50033E+00	.41704E+01	.25026E+01
3	8.8421	.41267E+01	.25146E+01	.40629E+00	.48363E+00	.41515E+01	-.25394E+01
4	9.5391	.19573E+01	.91631E+00	.19210E+01	.31236E+00	.29194E+01	.18784E+01
5	10.3032	.25970E+01	.15542E+01	.11983E+01	.31282E+00	.29181E+01	-.18753E+01
6	10.6460	.29026E+01	.18616E+01	.28375E+00	.31229E+00	.29194E+01	.18784E+01
7	11.1554	.17666E+01	.23497E+01	.13304E+01	.45948E+00	.24306E+01	.89897E+00
8	12.0490	.21956E+01	.70092E+00	.83613E+01	.44840E+00	.24196E+01	-.92496E+00
9	12.4499	.24183E+01	.88672E+01	.20147E+01	.45949E+00	.24306E+01	.89896E+00

#	
1	-26.5117
2	-14.9998
3	-3.4878
4	-26.6031
5	-14.9998
6	-3.3965
7	-26.5240
8	-14.9998
9	-3.4756

ELEMENT NO. = 9

#							
1	12.3313	.15514E+01	.10951E+01	.11046E+01	.43255E+00	.21029E+01	.66105E+00
2	13.3190	.19246E+01	.46305E+01	.68925E+00	.43846E+00	.21093E+01	-.64773E+00
3	13.7622	.20928E+01	.65094E+00	.16684E+01	.43255E+00	.21029E+01	.66105E+00
4	14.7558	.12088E+01	.10013E-01	.80156E+00	.36565E+00	.16103E+01	-.39148E+00
5	15.9378	.14751E+01	.25749E+00	.11831E+01	.36528E+00	.16091E+01	.39151E+00
6	16.4681	.16033E+01	.38446E+01	.50015E+00	.36565E+00	.16103E+01	.39148E+00
7	17.1803	.11286E+01	.28302E+00	.56163E+00	.42350E+00	.14088E+01	-.28421E-02
8	18.5565	.13094E+01	.85840E-01	.35322E+01	.41858E+00	.14041E+01	-.88043E-02
9	19.1739	.14036E+01	.80509E-02	.85358E+00	.42350E+00	.14088E+01	.28495E-02

#	
1	-26.5326
2	-14.9999
3	-3.4671
4	-26.6055
5	-15.0000
6	-3.3944
7	-26.5132
8	-15.0000
9	-3.4871

LOAD CASE # 2

THERMAL LOADING
   Ø Ø 1

 Ø

PRESCRIBED NODAL TEMPERATURES

1	25.000
2	33.333
3	41.667
4	53.333
5	65.000
6	82.500
7	100.000
8	25.000
9	41.667
10	65.000
11	100.000
12	25.000
13	33.333
14	41.667
15	53.333
16	65.000
17	82.500
18	100.000
19	25.000
20	41.667
21	65.000
22	100.000
23	25.000
24	33.333
25	41.667
26	53.333
27	65.000
28	82.500
29	100.000
30	25.000
31	41.667
32	65.000
33	100.000
34	25.000
35	33.333
36	41.667
37	53.333
38	65.000
39	82.500
40	100.000

TOTAL NODAL FORCES FOR EACH ELEMENT

Elem.						
1	-.1554E+02	-.1935E+03	.8771E+02	.5342E+02	.2392E+03	.6409E+02
	-.4925E+02	-.1439E+03	.2882E+02	.1883E+02	.1466E+03	.3928E+02
2	-.5010E+02	.4335E+03	.2117E+03	.1164E+03	.5929E+03	.1589E+03
	-.1252E+03	.3223E+03	.7945E+02	.3741E+03	.3855E+03	.1033E+03
3	.2991E+03	.1006E+04	.5003E+03	.2684E+03	.1409E+04	.3775E+03
	-.2991E+03	.7480E+03	.1929E+03	.8376E+02	.9278E+03	.2486E+03
4	-.1900E+01	.1913E+03	.1027E+03	.2411E+01	.1751E+03	.1751E+03
	-.2411E+01	.5573E+02	.3438E+02	.1900E+01	.1073E+03	.1073E+03
5	-.7329E+01	.4285E+03	.2416E+03	.5106E+01	.4340E+03	.4340E+03
	-.5106E+01	.1248E+03	.8751E+02	.7329E+01	.2822E+03	.2822E+03
6	.2391E+02	.9942E+03	.5674E+03	.1773E+02	.1031E+04	.1031E+04
	-.1773E+02	.2896E+03	.2089E+03	.2391E+02	.6792E+03	.6792E+03
7	-.1883E+02	.1378E+03	.9012E+02	.4925E+02	.6409E+02	.2392E+03
	-.5342E+02	.4739E+02	.3072E+02	.1554E+02	.3928E+02	.1466E+03
8	-.3741E+02	.3087E+03	.2066E+03	.1252E+03	.1589E+03	.5929E+03
	-.1164E+03	.1061E+03	.7212E+02	.5010E+03	.1033E+03	.3855E+04
9	-.8376E+02	.7162E+03	.4825E+03	.2991E+03	.3775E+03	.1409E+04
	.2684E+03	.2463E+04	.1690E+03	.1252E+03	.2486E+03	.9278E+03

DISPLACEMENTS

NODE	X-DISP.	Y-DISP.
1	.454700E+00	0
2	.496155E+00	0
3	.580163E+00	0
4	.745274E+00	0
5	.956658E+00	0
6	.134623E+01	0
7	.182053E+01	0
8	.438767E+00	.117570E+00
9	.559941E+00	.150034E+00
10	.923264E+00	.247385E+00
11	.175750E+01	.470923E+00
12	.393781E+00	.227353E+00
13	.429684E+00	.240080E+00
14	.502439E+00	.290085E+00
15	.645427E+00	.372639E+00
16	.828486E+00	.473328E+00
17	.116587E+01	.673117E+00
18	.157662E+01	.910265E+00
19	.321201E+00	.321201E+00

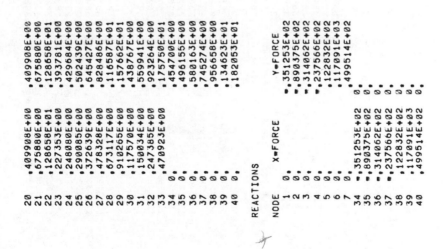

STRESSES

ELEMENT NO. = 1

G.P.	X-COORD.	Y-COORD.	X-STRESS	Y-STRESS	XY-STRESS	Z-STRESS	MAX P.S.	MIN P.S.	ANGLE
1	5.3653	.3219	.67792E+01	.58293E+02	.31676E+01	.32961E+02	.58487E+02	.65851E+01	3.5055
2	5.1925	1.3913	.10611E+02	.55312E+02	-.12904E+02	.33216E+02	.58700E+02	.71535E+01	15.0000
3	4.8074	2.4039	.16915E+02	.48157E+02	-.20722E+02	.32961E+02	.58486E+02	.65852E+01	26.4949
4	6.6538	.3992	.10388E+02	.38536E+02	.16171E+01	.31344E+02	.38629E+02	.10295E+02	3.2772
5	6.4395	1.7255	.12160E+02	.36669E+02	-.70749E+01	.31315E+02	.38564E+02	.10264E+02	14.9999
6	5.9619	2.9812	.16024E+02	.32899E+02	-.11380E+02	.31344E+02	.38628E+02	.10295E+02	26.7225
7	7.9423	.4765	.17910E+02	.27732E+02	-.66620E+00	.33587E+02	.27777E+02	.17865E+02	3.8627
8	7.6865	2.0596	.18011E+02	.26887E+02	-.25622E+01	.33364E+02	.27574E+02	.17326E+02	14.9995
9	7.1165	3.5585	.19788E+02	.25853E+02	-.39197E+01	.33587E+02	.27777E+02	.17865E+02	26.1356

ELEMENT NO. = 2

G.P.	X-COORD.	Y-COORD.	X-STRESS	Y-STRESS	XY-STRESS	Z-STRESS	MAX P.S.	MIN P.S.	ANGLE
1	8.8421	.5305	.17362E+02	.20190E+02	.20658E+00	.33414E+02	.20205E+02	.17347E+02	4.1566
2	8.5574	2.2929	.17909E+02	.20218E+02	-.66672E+00	.33586E+02	.20397E+02	.17730E+02	15.0000
3	7.9228	3.9617	.17890E+02	.19662E+02	-.11211E+01	.33414E+02	.20205E+02	.17347E+02	25.8426
4	10.6460	.6387	.15516E+02	.73038E+01	.52445E+00	.33513E+02	.15549E+02	.72704E+01	3.6399
5	10.3032	2.7607	.14926E+02	.77764E+01	-.20639E+01	.33477E+02	.15479E+02	.72234E+01	14.9995
6	9.5391	4.7699	.13917E+02	.89025E+01	-.32936E+01	.33513E+02	.15549E+02	.72705E+01	26.3592
7	12.4499	.7469	.14817E+02	.29996E+01	.97714E+01	.34730E+02	.14870E+02	.30531E+01	3.1299
8	12.0490	3.2285	.13226E+02	.20235E+01	.44021E+01	.34546E+02	.14406E+02	-.32031E+01	14.9997
9	11.1554	5.5781	.11209E+02	.60812E+01	.72260E+01	.34730E+02	.14870E+02	.30536E+01	26.8695

ELEMENT NO. = 3

G.P.	X-COORD.	Y-COORD.	X-STRESS	Y-STRESS	XY-STRESS	Z-STRESS	MAX P.S.	MIN P.S.	ANGLE
1	13.7622	.8256	.12147E+02	-.10548E+02	.13505E+01	.34952E+02	.12227E+02	-.10628E+02	3.3930
2	13.3190	3.5688	.11190E+02	-.88347E+01	.57806E+01	.35179E+02	.12739E+02	-.10384E+02	15.0000
3	12.3313	6.1661	.76429E+01	-.60435E+01	.91520E+01	.34952E+02	.12227E+02	-.10628E+02	26.6067
4	16.4681	.9880	.68310E+01	.25051E+02	.19999E+01	.35784E+02	.69560E+01	.25176E+02	3.5754
5	15.9378	4.2705	.47566E+01	.23083E+02	.80366E+01	.35752E+02	.69100E+01	.25236E+02	15.0000
6	14.7558	7.3784	.59225E+00	.18813E+02	.12806E+02	.35784E+02	.69558E+01	.25176E+02	26.4247
7	19.1739	1.1503	.19614E+01	.38612E+02	.23868E+01	.37033E+02	.21013E+02	.38752E+02	3.3551
8	18.5565	4.9722	-.11455E+01	.36237E+02	.10130E+02	.36813E+02	.15688E+01	.38551E+02	15.0000
9	17.1803	8.5908	.61152E+01	-.30535E+02	.16375E+02	.37033E+02	.21011E+02	.38752E+02	26.6451

ELEMENT NO.= 4

#								
1	4.4855	2.9614	.22401E+02	.42670E+02	-.23889E+02	.58486E+02	-.65850E+01	33.5057
2	3.8011	3.8011	.32961E+02	.32961E+02	-.25808E+02	.58769E+02	-.71530E+01	45.0000
3	2.9614	4.4855	.42670E+02	.22401E+02	-.23889E+02	.58486E+02	-.65850E+01	-33.5057
4	3.6726	5.5627	.18825E+02	.30098E+02	-.12997E+02	.38628E+02	-.10295E+02	33.2775
5	4.7141	4.7141	.24414E+02	.24414E+02	-.14150E+02	.38564E+02	-.10264E+02	45.0000
6	5.5627	3.6726	.30098E+02	.18825E+02	-.12997E+02	.38628E+02	-.10295E+02	-33.2775
7	6.6400	4.3838	.20942E+02	.22449E+02	-.24699E+02	.27776E+02	-.17865E+02	33.8622
8	4.3838	5.6270	.24699E+02	.24699E+02	-.20942E+02	.27573E+02	-.17325E+02	45.0000
9	5.6270	6.6400	.22449E+02	.20942E+02	-.24699E+02	.27776E+02	-.17865E+02	-33.8622

ELEMENT NO.= 5

#								
1	7.3923	4.8805	.18248E+02	.19304E+02	-.13275E+02	.20205E+02	-.17347E+02	34.1550
2	6.2645	6.2645	.19064E+02	.19064E+02	-.13331E+02	.20397E+02	-.17730E+02	45.0000
3	4.8805	7.3923	.19304E+02	.18248E+02	-.13275E+02	.20205E+02	-.17347E+02	-34.1550
4	5.8762	8.9004	.13009E+02	.98113E+01	-.38184E+01	.15549E+02	-.72703E+01	33.6414
5	7.5425	7.5425	.11351E+02	.11351E+02	-.41281E+01	.15479E+02	-.72232E+01	-45.0000
6	5.8762	5.8762	.98113E+01	.13009E+02	-.38184E+01	.15549E+02	-.72703E+01	-33.6414
7	10.4085	6.8718	.95163E+01	.23009E+02	-.23009E+02	.14870E+02	-.30530E+02	33.1304
8	8.8205	8.8205	.56013E+01	.56013E+01	-.88045E+01	.14406E+02	-.32033E+02	45.0000
9	6.8718	10.4085	.23009E+02	.95163E+01	-.82034E+01	.14870E+02	-.30530E+02	-33.1304

ELEMENT NO.= 6

#								
1	11.5056	7.5962	.53040E+01	-.37047E+01	.10502E+02	.34952E+02	-.10628E+02	33.3929
2	9.7502	9.7502	.11774E+02	.11774E+02	-.11561E+02	.35179E+02	-.10384E+02	-45.0000
3	7.5962	11.5056	.37047E+01	.53040E+01	-.10502E+02	.34952E+02	-.10628E+02	-33.3929
4	13.7678	11.0897	.28716E+02	-.15349E+02	.14805E+02	.35784E+02	-.25177E+02	33.5752
5	11.6673	11.6673	.91633E+01	.91633E+01	.16073E+02	.35752E+02	-.25237E+02	45.0000
6	9.0897	13.7678	.15349E+02	.28716E+02	-.14805E+02	.35784E+02	-.25177E+02	-33.5752
7	16.0299	10.5832	.10249E+02	-.18691E+02	.18762E+02	.37032E+02	-.21011E+02	33.3551
8	13.5843	13.5843	.18691E+02	.10249E+02	-.20260E+02	.36813E+02	-.15692E+02	45.0000
9	10.5832	16.0299	.26402E+02	-.10249E+02	.18762E+02	.37032E+02	-.21011E+02	-33.3551

ELEMENT NO.= 7

No.									
1	2.4039	4.8074	.48157E+02	.16915E+02	-.20722E+02	.32961E+02	.58486E+02	.65852E+01	-26.4949
2	1.3913	5.1925	.55312E+02	.10611E+02	.12904E+02	.33216E+02	.58770E+02	.71535E+01	-15.0002
3	.3219	5.3653	.58293E+02	.67792E+01	-.31676E+01	.32961E+02	.54487E+02	.65851E+01	-3.5055
4	2.9812	5.9619	.32899E+02	.16024E+02	-.11380E+02	.31344E+02	.36628E+02	.10295E+02	-26.7225
5	1.7255	6.4395	.36669E+02	.12160E+02	.70749E+01	.31315E+02	.38564E+02	.10264E+02	-14.9999
6	.3992	6.6538	.38536E+02	.10388E+02	.16171E+01	.31344E+02	.36629E+02	.10295E+02	-3.2772
7	3.5585	7.1165	.25853E+02	.19788E+02	.39197E+01	.33587E+02	.27777E+02	.17865E+02	-26.1356
8	2.0596	7.6865	.26887E+02	.18011E+02	-.25622E+01	.33364E+02	.27574E+02	.17325E+02	-14.9995
9	.4765	7.9423	.27732E+02	.17910E+02	.66620E+00	.33587E+02	.27777E+02	.17865E+02	-3.8627

ELEMENT NO.= 8

No.									
1	3.9617	7.9228	.19662E+02	.17890E+02	.11211E+01	.33414E+02	.20205E+02	.17347E+02	-25.8426
2	2.2929	8.5574	.20218E+02	.17909E+02	-.66672E+00	.33586E+02	.20397E+02	.17730E+02	-15.0000
3	.5305	8.8421	.20190E+02	.17362E+02	.20658E+00	.33414E+02	.20205E+02	.17347E+02	-4.1566
4	4.7699	9.5391	.89025E+01	.13917E+02	-.32936E+01	.35513E+02	.15549E+02	.72705E+01	-26.3592
5	2.7607	10.3032	.77764E+01	.14926E+02	.20639E+02	.33477E+02	.15479E+02	.72234E+01	-14.9995
6	.6387	10.6460	.73038E+01	.15516E+02	.52454E+00	.33513E+02	.15549E+02	.72704E+01	-3.6399
7	5.5781	11.1554	.60812E+00	.11209E+02	.72260E+01	.34730E+02	.14870E+02	.30536E+01	-26.8695
8	3.2285	12.0490	-.20235E+01	.13226E+02	.44021E+01	.34546E+02	.14406E+02	.32031E+01	-14.9997
9	.7469	12.4499	-.29996E+01	.14817E+02	.97714E+00	.34730E+02	.14870E+02	.30531E+01	-3.1299

ELEMENT NO.= 9

No.									
1	6.1661	12.3313	-.60435E+01	.76429E+01	.91520E+01	.34952E+02	.12227E+02	.10628E+02	-26.6067
2	3.5688	13.3190	-.88347E+01	.11190E+02	.57806E+01	.35179E+02	.12739E+02	.10384E+02	-15.0000
3	.8256	13.7622	.10548E+02	.12147E+02	.13503E+01	.34952E+02	.12227E+02	.10628E+02	-3.3930
4	7.3784	14.7558	.18813E+02	.59225E+00	.12806E+02	.35784E+02	.69558E+01	.25176E+02	-26.4247
5	4.2705	15.9378	.23083E+02	.47566E+01	.80366E+02	.35752E+02	.69100E+01	.25236E+02	-15.0000
6	.9880	16.4681	.25051E+02	.68310E+01	.19999E+01	.35784E+02	.69560E+01	.25176E+02	-3.5754
7	8.5908	17.1803	.30535E+02	.61152E+01	.16375E+02	.37033E+02	.21011E+01	.38752E+02	-26.6451
8	4.9722	18.5565	.36237E+02	.11455E+02	.10130E+02	.36813E+02	.15688E+01	.38951E+02	-15.0000
9	1.1503	19.1739	-.38612E+02	.19614E+01	.23868E+01	.37033E+02	.21013E+01	.38752E+02	-3.3551

subdivision employed in solution is also illustrated and it is seen that 9 parabolic isoparametric elements have been utilised.

The first loading case considered is the application of an internal pressure of 10 units, with the external boundary being unloaded. Both the radial and hoop stress distributions are compared with the theoretical values [1] in Figs 10.5(a) and (b) respectively where good agreement is evident in spite of the relatively coarse element mesh employed. The results obtained by use of both two- and three-point numerical integration rules are shown.

The second loading case assumed is the thermal condition when a linear radial temperature variation exists in the cylinder; the temperature values assumed being indicated in Fig. 10.4. When this temperature distribution is input into the program, the resulting thermal stresses can be found. Mechanically, both inside and outside surfaces are assumed to be free from tractions. The circumferential thermal stress distribution is compared with the theoretical values [2] in Fig. 10.6 where excellent agreement is evident. Comparison of the radial displacement values indicates errors in the range 0·1–0·5%.

10.4.3 Plate bending applications

The first plate example is the simply supported square plate shown in Fig. 10.7 where the plate dimensions and the material properties assumed

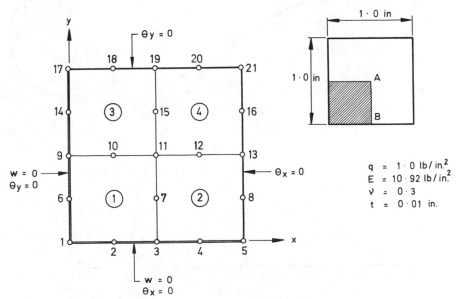

FIG. 10.7.   Finite element mesh in symmetric quadrant of simply supported square plate.

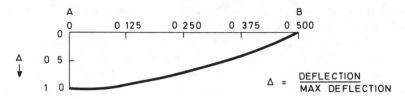

FIG. 10.8. Deflection of a square plate with simply supported edges.

are also described. The plate is subjected to a uniformly distributed loading, $q = 1.0$ and the problem is solved using 4 parabolic isoparametric elements in the symmetric quadrant as illustrated in Fig. 10.7. The lateral displacement distribution along the centreline is shown in Fig. 10.8 and compares well with a solution based on thin plate theory [3]. The error in the central displacement is 1 %. The stress resultants are also in good agreement with the thin plate solution with typical errors of less than 5 %.

Details of the second example are given in Fig. 10.9. This plate which is

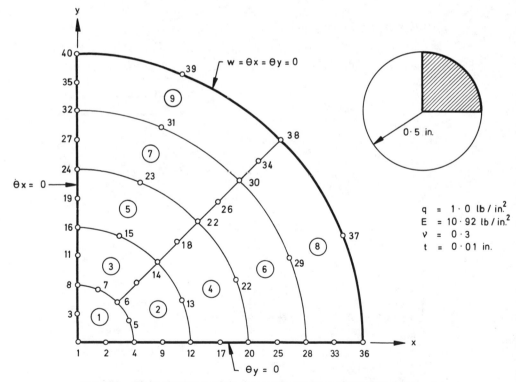

FIG. 10.9. Finite element mesh in symmetric quadrant of clamped circular plate

OUTPUT FOR PLATE BENDING TEST EXAMPLE

TOTAL NO. OF PROBLEMS = 1

PROBLEM NO. 1   ** CLAMPED CIRCULAR PLATE UNDER UDL **

NPOIN = 40   NELEM = 9   NVFIX = 24   NCASE = 1   NTYPE = 0   NNODE = 8   NDOFN = 3

NMATS = 1   NPROP = 4   NGAUS = 2   NDIME = 2   NSTRE = 5   NEVAB = 24

ELEMENT   PROPERTY   NODE NUMBERS

ELEMENT	PROPERTY								
1	1	12	17	20	21	22	18	14	13
2	1	14	18	22	23	24	19	16	15
3	1	1	2	4	5	6	7	8	3
4	1	20	25	28	29	30	26	22	21
5	1	4	9	12	13	14	10	6	5
6	1	6	10	14	15	16	11	8	7
7	1	22	26	30	31	32	27	24	23
8	1	28	33	36	37	38	34	30	29
9	1	30	34	38	39	40	35	32	31

NODAL POINT COORDINATES

NODE	X	Y
1	0.000	0.000
2	.050	0.000
3	0.000	.050
4	.100	0.000
5	.092	.038
6	.071	.071
7	.038	.092
8	0.000	.100
9	.150	0.000
10	.106	.106
11	0.000	.150

Node		
12	.200	0.000
13	.185	.077
14	.141	-.141
15	.077	-.185
16	0.000	-.200
17	.250	-.177
18	.177	-.250
19	0.000	-.300
20	.300	-.115
21	.277	-.212
22	.212	-.277
23	.115	-.300
24	0.000	-.350
25	.350	-.247
26	.247	-.350
27	0.000	-.400
28	.400	-.153
29	.370	-.283
30	.283	-.370
31	.153	-.400
32	0.000	-.450
33	.450	-.318
34	.318	-.450
35	0.000	-.500
36	.500	-.191
37	.462	-.354
38	.354	-.462
39	.191	-.500
40	0.000	-.500

RESTRAINED NODES

NODE	CODE		FIXED VALUES	
1	011	-0.000000	-0.000000	-0.000000
2	001	-0.000000	-0.000000	-0.000000
3	010	-0.000000	-0.000000	-0.000000
4	001	-0.000000	-0.000000	-0.000000
8	010	-0.000000	-0.000000	-0.000000
9	001	-0.000000	-0.000000	-0.000000
11	010	-0.000000	-0.000000	-0.000000
12	001	-0.000000	-0.000000	-0.000000
16	010	-0.000000	-0.000000	-0.000000

```
17  001  -0.000000  -0.000000  -0.000000  -0.000000  -0.000000
19  010  -0.000000  -0.000000  -0.000000  -0.000000  -0.000000
20  001  -0.000000  -0.000000  -0.000000  -0.000000  -0.000000
24  010  -0.000000  -0.000000  -0.000000  -0.000000  -0.000000
25  001  -0.000000  -0.000000  -0.000000  -0.000000  -0.000000
27  010  -0.000000  -0.000000  -0.000000  -0.000000  -0.000000
28  001  -0.000000  -0.000000  -0.000000  -0.000000  -0.000000
32  010  -0.000000  -0.000000  -0.000000  -0.000000  -0.000000
33  001  -0.000000  -0.000000  -0.000000  -0.000000  -0.000000
35  010  -0.000000  -0.000000  -0.000000  -0.000000  -0.000000
36  111  -0.000000  -0.000000  -0.000000  -0.000000  -0.000000
37  111  -0.000000  -0.000000  -0.000000  -0.000000  -0.000000
38  111  -0.000000  -0.000000  -0.000000  -0.000000  -0.000000
39  111  -0.000000  -0.000000  -0.000000  -0.000000  -0.000000
40  111  -0.000000  -0.000000  -0.000000  -0.000000  -0.000000
```

MATERIAL PROPERTIES
NUMBER
   PROPERTIES   .109200E+02   .300000E+00   .100000E+01   .100000E+01

MAXIMUM FRONTWIDTH ENCOUNTERED =   60

** UNIFORMLY DISTRIBUTED LOAD **
0

TOTAL NODAL FORCES FOR EACH ELEMENT   LOAD CASE = 1

```
1   -.1744E-02   0.          .6540E-02   0.         -.1526E-02   0.          .6540E-02
     0.          .6976E-02    0.          .6104E-02   0.          .1526E-02    0.

2   -.1744E-02   0.         -.1744E-02   0.          .1526E-02   0.          .6540E-02
     0.          .6976E-02    0.          .6104E-02   0.          .1526E-02    0.

3   -.5697E-03   0.          .2360E-02   0.          .2872E-02   0.          .2360E-02
     0.         -.6104E-03   0.         -.8255E-03   0.         -.6104E-03    0.

4    .2398E-02   0.          .9592E-02   0.          .9156E-02   0.          .9156E-02
     0.         -.2398E-02   0.          .2180E-02   0.          .2180E-02    0.
```

NODE	DISP.	XZ-ROT.	YZ-ROT.
1	.966456E+03	-0.	-0.
2	.951865E+03	-.744799E+03	-.744799E+03
3	.951865E+03	-0.	-0.
4	.895522E+03	-.147117E+04	.594824E+03
5	.896087E+03	-.132280E+04	.106030E+04
6	.894720E+03	-.106030E+04	.132280E+04
7	.896087E+03	-.594824E+03	.147117E+04
8	.895522E+03	-0.	-0.
9	.806769E+03	-.211635E+04	.147724E+04
10	.805173E+03	-.147724E+04	.211635E+04
11	.806769E+03	-0.	-0.
12	.688447E+03	-.256376E+04	.101302E+04
13	.686722E+03	-.243042E+04	.180329E+04
14	.687976E+03	-.180329E+04	.243042E+04
15	.686722E+03	-.101302E+04	.256376E+04
16	.688447E+03	-0.	-0.
17	.548806E+03	-.291061E+04	.205064E+04
18	.548801E+03	-.205064E+04	.291061E+04
19	.548806E+03	-0.	-0.
20	.399657E+03	-.300348E+04	.113790E+04
21	.399406E+03	-.275482E+04	.212316E+04
22	.399992E+03	-.212316E+04	.275482E+04
23	.399406E+03	-.113790E+04	.300348E+04
24	.399657E+03	-0.	-0.
25	.254205E+03	-.279002E+04	.197498E+04
26	.254529E+03	-.197498E+04	

27	.254205E+03	-0.	.223106E+04	-0.	.279002E+04
28	.127364E+03	-0.	.207915E+04	-0.	.859950E+03
29	.126427E+03	-0.	.157965E+04	-0.	.157965E+04
30	.127563E+03	-0.	.859950E+03	-0.	.207915E+04
31	.126427E+03	-0.	.223106E+04	-0.	.223106E+04
32	.127364E+03	-0.	.133741E+04	-0.	.133741E+04
33	.357009E+02	-0.	.947607E+03	-0.	.947607E+03
34	.357738E+02	-0.	.947607E+03	-0.	.947607E+03
35	.357009E+02	-0.	.133741E+04	-0.	.133741E+04
36	-0.	-0.	-0.	-0.	-0.
37	-0.	-0.	-0.	-0.	-0.
38	-0.	-0.	-0.	-0.	-0.
39	-0.	-0.	-0.	-0.	-0.
40	-0.	-0.	-0.	-0.	-0.

REACTIONS

NODE	FORCE	XZ-MOMENT	YZ-MOMENT
1	-0.	-.915065E-04	-.915065E-04
2	-0.	.182649E-02	.182649E-02
3	-0.	-0.	-0.
4	-0.	.544808E-03	.544808E-03
8	-0.	.707997E-03	.707997E-03
9	-0.	-.877845E-03	-.877845E-03
11	-0.	.833122E-03	.833122E-03
12	-0.	-0.	-0.
16	-0.	-.321331E-03	-.321331E-03
17	-0.	-0.	-0.
19	-0.	.382164E-03	.382164E-03
20	-0.	-0.	-0.
24	-0.	.639364E-05	.639364E-05
25	-0.	-0.	-0.
27	-0.	.160433E-03	.160433E-03
28	-0.	-0.	-0.
32	-0.	-0.	-0.
33	-0.	-0.	-0.
35	-0.	-0.	-0.
36	.152456E-01	.211141E-02	.664846E-04
37	-.675752E-01	.740127E-02	.306326E-02
38	-.305552E-01	.299130E-02	.299130E-02
39	-.675752E-01	.306326E-02	.740127E-02
40	-.152456E-01	.664846E-04	.211141E-02

STRESSES

G.P.	X-COORD.	Y-COORD.	X-MOMENT	Y-MOMENT	XY-MOMENT	XZ-S.FORCE	YZ-S.FORCE
ELEMENT NO. = 1							
1	.2178	.0373	.10427E-01	.14464E-01	.59401E-03	-.92355E-01	-.49524E-01
2	.1804	.1276	.11881E-01	.12840E-01	-.20134E-02	-.15118E+00	.25736E-02
3	.2747	.0471	.44555E-02	.10789E-01	-.12065E-02	-.14340E+00	-.82727E-02
4	.2275	.1609	.65408E-02	.88454E-02	.31590E-02	-.10035E+00	-.93391E-01
ELEMENT NO. = 2							
1	.1276	.1804	.12840E-01	.11881E-01	.20134E-02	.25736E-02	-.15118E+00
2	.0373	.2178	.14464E-01	.10427E-01	-.59401E-03	-.49524E-01	-.92355E-01
3	.1609	.2275	.88454E-02	.65408E-02	-.31590E-02	-.93391E-01	.10035E+00
4	.0471	.2747	.10789E-01	.44555E-02	.12065E-02	-.82727E-02	-.14340E+00
ELEMENT NO. = 3							
1	.0212	.0212	.19134E-01	.19134E-01	.70017E-04	.99380E-01	.99380E-01
2	.0188	.0781	.19313E-01	.18620E-01	-.12608E-03	.64275E-01	-.13901E-01
3	.0781	.0188	.18620E-01	.19313E-01	-.12608E-03	-.13901E-01	.64275E-01
4	.0659	.0659	.18892E-01	.18892E-01	.94134E-03	.16924E+00	.16924E+00
ELEMENT NO. = 4							
1	.3163	.0542	-.64716E-03	.77691E-02	-.15333E-02	.15686E+00	-.32685E-01
2	.2620	.1853	.20742E-02	.51745E-02	-.41982E-02	-.13019E+00	-.94417E-01
3	.3732	.0640	-.89370E-02	.29017E-02	-.20256E-02	-.17222E+00	-.24741E-01
4	.3091	.2186	-.50448E-02	-.96150E-03	-.59236E-02	-.15428E+00	.11071E+00

```
ELEMENT NO.= 5
   .1193   1  .0205  .18341E-01  .19207E-01 -.44524E-03 -.10742E+00  .66251E+00
   .0988   2  .0699  .17269E-01  .17275E-01 -.34347E-03  .54824E-01 -.29301E+00
   .1762   3  .0302  .13705E-01  .16565E-01 -.28353E-03 -.10134E+00 -.17254E+00
   .1459   4  .1032  .15000E-01  .14949E-01  .13414E-02  .32871E-01 -.40310E-01

ELEMENT NO.= 6
   .0699   1  .0988  .17275E-01  .17269E-01 -.34347E-03  .29301E+00  .54824E-01
   .0205   2  .1193  .19207E-01  .18341E-01 -.44524E-03  .66251E+00 -.10742E+00
   .1032   3  .1459  .14949E-01  .15000E-01  .13414E-02  .40310E-01 -.32871E-01
   .0302   4  .1762  .16565E-01  .13705E-01 -.28353E-03  .17254E+00 -.10134E+00

ELEMENT NO.= 7
   .1853   1  .2620  .51745E-02  .20742E-02 -.41982E-02  .94417E-01  .13019E+00
   .0542   2  .3163  .77691E-02  .64716E-03 -.15333E-02 -.32685E-01 -.15686E+01
   .2186   3  .3091  .96150E-03  .50448E-02 -.59236E-02 -.11071E+00 -.15428E+00
   .0640   4  .3732  .29017E-02  .89376E-02 -.20256E-02 -.24741E-01 -.18722E+00

ELEMENT NO.= 8
   .4148   1  .0711 -.15759E-01 -.11765E-02 -.25687E-02  .20752E+00 -.34445E-01
   .3436   2  .2430 -.11039E-01 -.58705E-02 -.72892E-02  .16920E+00  .12529E+00
   .4717   3  .0809 -.26360E-01 -.74717E-02 -.33080E-02  .23150E+00 -.64072E-01
   .3907   4  .2764 -.20251E-01 -.13617E-01 -.94533E-02  .21032E+00 -.11775E+00

ELEMENT NO.= 9
   .2430   1  .3436 -.58705E-02 -.11039E-01 -.72892E-02 -.12529E+00  .16920E+00
   .0711   2  .4148 -.11765E-02 -.15759E-01 -.25687E-02 -.34445E-01  .20752E+00
   .2764   3  .3907 -.13617E-01 -.20251E-01 -.94533E-02 -.11775E+00  .21032E+00
   .0809   4  .4717 -.74717E-02 -.26360E-01 -.33080E-02 -.64072E-01 -.23150E+00
```

circular in plan is subjected to a uniformly distributed loading $q = 1{\cdot}0$. The problem is solved using 9 parabolic isoparametric elements in a symmetric quadrant. The lateral displacement distribution along the centre-line is shown in Fig. 10.10 and is in good agreement with the theoretical solution.

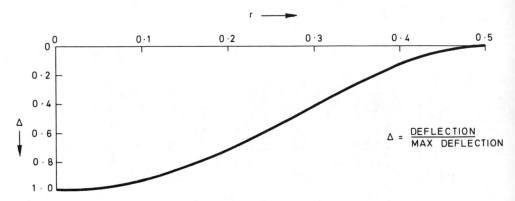

Fig. 10.10.   Deflection of a circular plate with clamped edges.

For example, the central displacement $w = 966{\cdot}456$ is less than 1% in error. The stress resultants are also in good agreement with the thin plate solution. Typically, the bending moment at $x = y = 0{\cdot}212$ is less than 5% in error.

References

1. Timoshenko, S., and Goodier, J. N., "Theory of Elasticity". McGraw-Hill, New York, 1951, p. 59.
2. Boley, B. A., and Weiner, J. H., "Theory of Thermal Stresses". Wiley, 1960, pp. 288–291.
3. Timoshenko, S., and Woinowsky-Krieger, S., "Plates and Shells". McGraw-Hill, New York, 1959.

11

Discussion and Further Developments

11.1 Introduction

In the preceding chapters of this text we have attempted to show how iso-parametric finite elements are programmed for structural applications. In particular we have discussed one and two dimensional parabolic elements illustrating the program organisation required for frontal equation solution. Finally the process of equation solution by the frontal technique has been itself described. As stated at the outset in Chapter 1 a compromise has had to be made between the development of commerically oriented programs and the presentation of readily understandable coding. Bearing in mind that the prime role of this text is a teaching one, the presentation is undoubtedly biased towards the latter.

The aim of this chapter is to review the programs presented and to indicate areas where improvements and savings can be made. (It should again be borne in mind, that the remarks in this chapter are largely aimed towards structural applications). Reductions in both core storage requirements and computer costs generally arise from three main sources:

Improvements in techniques and algorithms for computing the element stiffnesses

The development of more efficient solvers for large systems of equations. Savings in this area generally have the most influence on the core requirements and computer costs of a finite element solution.

Auxiliary improvements due to increased efficiency in data transfer to and from disc files, structure of common block arrays, etc. Savings in this area can be appreciable, but the optimum requirements may differ from machine to machine.

267

In addition to savings in solution costs brought about by improved coding, we must also consider the economies associated with the introduction of further facilities into the program. In this context, for example, the introduction of boundary restraints in directions not coincident with the coordinate axes, often allows a smaller segment of the structure to be analysed. On the other hand, the use of semi-analytical Fourier techniques [1,2] allows the non-symmetric loading of axisymmetric structures, to be solved as a series of two dimensional problems in place of a single three-dimensional analysis Of course, the cost savings or otherwise of this approach depends on the number of Fourier terms required to represent the loading accurately; but a reduction in core storage requirements always results.

The improvements envisaged for the programs presented in this text will now be individually discussed. Indeed many of them can be incorporated by the reader with only a little effort.

11.2 Element stiffness formulation

The stiffness matrix of any element is formed according to (1.28) of Chapter 1 and, as will be recalled, depends entirely on the strain **B** and elasticity **D** matrices. Since both matrices contain several zero entries, many arithmetic operations in the matrix multiplications in (1.28) are redundant. To achieve economies in this area kernel functions can be employed where preliminary integrations are performed and the element stiffnesses are obtained by simply multiplying by appropriate coefficients [3]. For example, for three-dimensional solid elements an integration array

$$W_{ij}^{mn} = \int_V \frac{\partial N_m}{\partial x_i} \cdot \frac{\partial N_n}{\partial x_j} \cdot dV, \tag{11.1}$$

is first evaluated, where m and n range over the element nodal connection numbers and i and j range from 1 to 3 so that x_1, x_2, x_3 denote x, y and z respectively. The sub-stiffness matrix linking nodes m and n is then given by

$$\mathbf{K}_{ij}^{mn} = \lambda W_{ij}^{mn} + \mu \left(W_{ij}^{mn} + \sum_{K=1}^{3} W_{KK}^{mn} \cdot \delta_{ij} \right), \tag{11.2}$$

where λ and μ are the Lamé constants ($\lambda = 2\mu v/(1 - 2v)$ and μ is the shear modulus) and δ_{ij} is the Kronecker delta; having zero value if $i \neq j$ and 1 if $i = j$). Matrix $\mathbf{K}_{ij}^{mn}$ is a 3×3 submatrix linking the three degrees of freedom of nodes m and n. Experience with this algorithm indicates that its use reduces element stiffness matrix computation times by approximately 40%.

An additional method of reducing the computer time required to formulate the element stiffnesses is to take advantage of the fact that element stiffness matrices are independent of coordinate translations or rotations. Thus, if a large number of elements with identical geometries are employed, the stiffness matrix need only be evaluated once. Of course, the degree of success of this artifice depends on the structural geometry and if a relatively small number of isoparametric elements are to be employed the savings may only be marginal. On the other hand if a large number of simple elements are employed in a three-dimensional situation, where stiffness formulation is relatively expensive, the benefit may be significant.

11.3 Multiple loading cases

If the structural configuration to be analysed is subjected to several loading conditions, then considerable savings can be obtained by processing all loading cases simultaneously. As a guide, solution for additional loading cases can be obtained in the order of 5–10 % of the time required for a single analysis. This economy is possible since a second or subsequent loading case requires only the formation of an additional load or right hand side term in the frontal solution. Since the arithmetic associated with the formation, reduction and eventual backsubstitution of the additional term is minimal, the increase in solution time is small. This task should be well within the capabilities of the reader, requiring only the conversion of GLOAD into a two-dimensional array GLOAD(IFRON,NORHS), where the first subscript refers to the frontal position, and the second relates to the right hand side or loading case number. Any operation performed on GLOAD in subroutine FRONT must then be carried out for each right hand side in turn.

11.4 Substructuring

The technique of substructuring generally comprises the following steps.

Assembly of the element matrices of a substructure to give the global stiffness matrix of the substructure.

Elimination of the terms associated with the internal nodal points (i.e. nodes not on the boundary) of the substructure. This process is known as *condensation*.

Solution of the system of resulting simultaneous equations obtained by assembling all the individual substructures. This gives the nodal displacements and reactions for all nodal points on interfaces between substructures.

Return to the individual substructures to evaluate the displacements at interior nodes and finall obtain the element stresses.

When computer core storage is at a premium, substructuring permits the solution of large problems. If the structure is divided into several substructures and each one is considered in turn, the problem can be reduced to a manageable size. Further, substructuring allows economies when it is required to perform several analyses in cases where the geometry of a local region only is changed. By considering the major portion of the solid as one substructure and the local changing region as another, it is only necessary to regenerate the global stiffness of the latter region for any reanalysis.

The very nature of the frontal method makes the use of substructure techniques a simple affair, since, when the front has advanced into a structure to a certain position, the reduced frontal equations are essentially the condensed equations for a substructure corresponding to the part of the structure already considered.

11.5 Temporary storage of the reduced equations

In the frontal subroutine presented, once a variable is eliminated the corresponding reduced equation is immediately transferred to a backing disc file. The sequence is followed in reverse for the backsubstitution. While this aids the understanding of the frontal technique, from a practical point of view it makes the process inefficient, since the transfer of information to and from disc files is a relatively slow operation.

This situation can be improved by storing the reduced equations corresponding to eliminated variables in core in a temporary array normally termed a *buffer* area. As soon as this array is full, the information is then transferred to disc. The number of equations which can be accommodated in this buffer area is governed by the buffer length, an integer quantity to be specified in the program. Thus on elimination of a variable a counter over the number of eliminated variables is incremented by one and the reduced equation is stored in core. The counter is checked against the permissible buffer length. If this has been reached, the buffer array is transferred to disc file and the counter reset to zero. On the backsubstitution the contents of a complete buffer length are read from disc file by backspacing. Backsubstitution for all the variables contained in the buffer is undertaken before a further record is read. Incorporation of such a facility is essential for efficient solution and is a task which should be within the capabilities of the reader.

11.6 Variable number of degrees of freedom per node

A condition which is more difficult to accommodate in the frontal solution is the situation where a different number of degrees of freedom exist at each node, extensive reprogramming is necessary to accommodate this. Unfortunately this facility is demanded by some elements, such as the Semiloof thin shell element [4]. In the parabolic isoparametric version of this element, each corner node has 3 degrees of freedom while each midside node has 5 associated degrees of freedom.

11.7 Multiple elements

A situation which can be easily accommodated is the case when more than one element type is employed to simulate the structure. For example the 8 noded two-dimensional plane stress/strain element may have a 3 noded line element associated with it. This is particularly useful in the analysis of reinforced concrete structures and prestressed concrete reactor vessels, where the concrete structure is modelled by solid elements and the steel pre-stressing systems and liner are idealised by the use of membrane elements which carry only in-plane loads [5].

Provided that the same number of degrees of freedom exist at each nodal point of each element, the present solution package (Chapter 8, Section 8.9.4) can be readily adapted to deal with this situation. It is first necessary to set NNODE in the input data equal to the largest value to be encountered. Since array LNODS must be dimensioned to accommodate the element with the greatest number of nodes, the array will contain zero entries for elements with fewer nodes. Thus typical entries for an eight and three noded element would, respectively, be

$$27 \quad 28 \quad 29 \quad 38 \quad 45 \quad 47 \quad 46 \quad 39$$
$$45 \quad 47 \quad 46 \quad 0 \quad 0 \quad 0 \quad 0 \quad 0$$

Directly after the element stiffnesses are read from file in subroutine FRONT the array LOCEL must be set to zero as follows

DO 165 IEVAB = 1, NEVAB

165 LOCEL(IEVAB) = 0

This is necessary since NEVAB will be based on the element with the largest number of degrees of freedom and values will not be assigned to components of the LOCEL vector corresponding to zero entries in the LNODS array for elements with fewer degrees of freedom.

The second change required is to be made directly after the definition of the quantity NIKNO in the DO LOOP to index 210.

IF (NIKNO.EQ.0) GO TO 210

This ensures that a zero nodal value in the element connection array is not assigned a position in the destination vector.

Finally, directly after entering the DO LOOP to index 240 the following statement must be inserted.

IF (LOCEL(IEVAB).EQ.0) GO TO 240

This ensure that entries are not made in the load and stiffness arrays corresponding to zero nodal values in the nodal connection array.

Of course it will also be necessary to incorporate additional subroutines for stiffness formulation, load vector construction, etc., elsewhere in the program.

11.8 Equation re-solution facility

If any of the programs discussed are to be extended to the realm of non-linear behaviour the resolution facility included in the frontal solver as described in Chapter 8, is essential. Figure 11.1 shows schematically the "residual force method" of solution for non-linear problems. In this, solution of the non-linear problem is achieved by performing a series of linear elastic analyses, attempts being made to satisfy the non-linear constitutive law at each stage. The force difference between the applied forces and the stresses satisfying the constitutive relation is interpreted as a residual force system which must be treated as additional nodal forces and eliminated by iteration.

Consequently such an analysis requires a series of linear elastic solutions. If the "initial stress" approach [6] is adopted as schematically illustrated in Fig. 11.1(a) several solutions employing the same element stiffnesses must be performed, and considerable economy in computer costs can be achieved if an equation resolution facility is available. Generally, a resolution can be performed in approximately 10–15% of the time for a complete solution.

11.9 Non-symmetric matrices

In some finite element applications non-symmetric matrices are unavoidably encountered. For example in the non-linear analysis of quasi-harmonic problems by the Newton–Raphson technique [7] the derivative matrix may be unsymmetric. The frontal subroutine FRONT (Chapter 8, Section

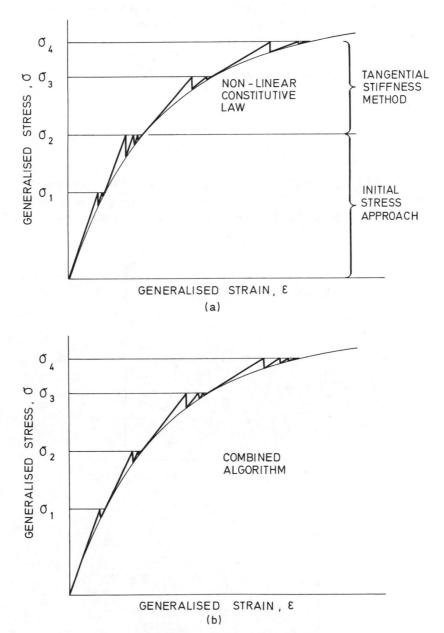

FIG. 11.1. Typical non-linear solution algorithms.

8.9.4) can be easily adapted to deal with unsymmetric stiffnesses, at the expense, of course, of an increase in core storage requirements.

In this case, it is necessary to store the complete global stiffness submatrix, and not just the upper triangular portion. Thus the main modification required is the conversion of the global stiffness matrix to the two-dimensional array form GSTIF(IFRON,JFRON). This task is a further one which can readily be undertaken by the reader.

11.10 Inclined boundary axes

As previously mentioned, the use of boundary axes which are not aligned with the coordinate axes often allows considerable savings in both computer costs and data preparation. An example of this is illustrated in Fig. 11.2.

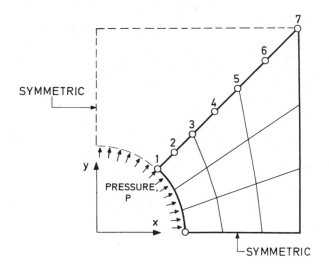

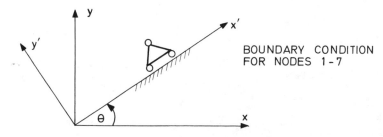

FIG. 11.2.   Locally directed boundary conditions.

Once again, this facility can be included without too much difficulty. Suppose, for example, it is required to restrict displacements to directions along either the x' or y' axes inclined at an angle θ to the x, y axes as shown in Fig. 11.2. The first step is to transform the nodal displacement and force components to the x', y' directions. For the plane stress/strain application this is accomplished by use of the transformation matrix, $\mathbf{T}$, so that

$$\begin{bmatrix} P_x \\ P_y \end{bmatrix}' = [\mathbf{T}]^{\mathrm{T}} \begin{bmatrix} P_x \\ P_y \end{bmatrix}; \qquad \begin{bmatrix} u \\ v \end{bmatrix} = \mathbf{T} \begin{bmatrix} u \\ v \end{bmatrix}' \tag{11.3}$$

where $\mathbf{T}$ is given, for this case, by

$$\mathbf{T} = \begin{bmatrix} \cos\theta & -\sin\theta \\ \sin\theta & \cos\theta \end{bmatrix} \tag{11.4}$$

Since the nodal forces and displacement are linked through the stiffness matrix by

$$\begin{bmatrix} P_{xi} \\ P_{yi} \end{bmatrix} = \mathbf{K}_{ij} \begin{bmatrix} u_j \\ v_j \end{bmatrix} \tag{11.5}$$

where $\mathbf{K}_{ij}$ is the 2×2 submatrix, the equivalent expression with respect to the primed coordinate system is given on substitution of (11.3) in (11.5) as

$$\begin{bmatrix} P_{xi} \\ P_{yi} \end{bmatrix}' = \mathbf{K}'_{ij} \begin{bmatrix} u_j \\ v_j \end{bmatrix}' \tag{11.6}$$

where

$$\mathbf{K}'_{ij} = [\mathbf{T}]^{\mathrm{T}} \mathbf{K}_{ij} \mathbf{T} \tag{11.7}$$

Thus the nodal forces and displacements associated with each variable which is to be restrained with respect to local directions must be transformed so as to be coincident with the local axes. The steps to be taken in solution can be summarised as follows:

For any node to be restrained in local directions, transform the applied nodal forces to coincide with the local axes.

Also transform the relevant element stiffness submatrices according to (11.7).

Assemble the loads and stiffnesses in the usual way and solve for the displacements. In transforming the load and element stiffnesses to local directions it is understood that the corresponding displacements are also referred to the local directions. Hence on solution all displacements are with respect to the global axes, x, y except for the locally restrained variables which will be automatically computed in the local directions.

Transform the displacements and reactions in the local directions to the global reference frame before evaluating the stresses.

11.11 Sectorial symmetry

A situation which is often encountered in engineering is that in which the geometry and loading repeat regularly, but the boundaries of the repeated segments are not lines of symmetry. The most common examples of this are pump impellors or fans with aerofoil blades. Such a situation is shown in Fig. 1.1 where the complete fan is composed of 9 of the segments illustrated. Displacement conditions at the segment boundaries are not known absolutely. However it is known that the displacements of a particular node on one boundary (marked m in Fig. 1.1) must be equal to those of the corresponding node marked s (in Fig. 1.1) on the other boundary.

The frontal method of solution allows such a situation to be treated easily. Since the nodal numbers are not prime quantities, being merely nicknames employed to identify the nodes within an element, it is the element numbering which is of importance. Therefore it is relatively easy to arrange that the stiffness contributions of two distinct nodal points be assembled into the same location in the global stiffness array. One of the nodes to be coupled in this way is designated the master node with the other being the slave node; this information being provided as input data for each such pair of nodes. The element stiffness contribution of the master node is then assembled in its usual location. However, the stiffness contribution of the slave node is assembled in the same location as the master node regardless of its nodal point number. In this way the desired coupling effect is easily achieved.

11.12 Stress analysis of three-dimensional solids and axisymmetric problems

An extension which the more advanced reader may carry out is the modification of the plane stress/strain program to allow the analysis of fully three-dimensional solids. The development of such a program would be particularly instructive since it is in the three dimensional situation that the advantages of the frontal equation solution technique are most marked. Three-dimensional problems generally have a high element connectivity and the assembly/elimination process of the frontal method ensures a more efficient solution of equations than if a banded matrix solver were employed.

Most of the necessary changes are obvious, with the general three-dimensional equations of elasticity being used in place of the reduced plane stress/strain versions. All quantities relating to coordinate dimensions now become 3 and the basic parabolic element is now a 20 noded one for which the shape functions can be found in [8].

On the other hand conversion of the plane stress/strain program to accommodate axisymmetric situations is a relatively simple task. The main change required is to the **D** and **B** matrices which now have an extra component corresponding to the circumferential (or hoop) stress and strain terms. Explicit forms for these matrices are given in [8]. Numerical integrations over element volumes and surfaces must also be amended to include the circumferential direction.

11.13 The need for dynamic dimensions and associated diagnostics

The diagnostics which relate to dimensioning are particularly important, because if an array dimension is exceeded a correct item of data is placed in another array, or lost through being over-written by another misplaced item of data: usually the job runs, but gives absurd answers. Such diagnostics are not easily provided in a thoroughly foolproof way, because standard FORTRAN lacks one simple basic facility, and this lack drives us to one of several strategic choices.

1. We can use COMMON statements including say COORD (200,3). Suppose NPOIN the total number of nodal points is then input as 204, which is too big. In order to discover this, the program must know that COORD allows only 200 rows. It would be easy if we could *ask* for the dimensions of COORD in the program. A possible solution is to introduce statement ICORD = 200, as near to the COMMON statement as possible, for example in a DATA statement with a number of other similar dimensions. It is then understood that if the COMMON statement is altered, the DATA statement must be altered too.
2. An improvement on the above approach is to lay out COMMON in some definite way, e.g.

    COMMON COORD(300,3), ELOAD(100,8)...

and then to equivalence this with some vector VECTO

    DIMENSION VECTO(10000)
    EQUIVALENCE (COORD(1,1), VECTO(1))...

Some arbitrary number, say 10^{30}, can then be put into ELOAD(1,1) etc. and we immediately check that 10^{30} appears in all the expected positions in VECTO; otherwise the user is penalised by stopping his program.

3. We can imitate machine code. The COMMON reduces to VECTO(10000) and the starting location of say COORD is stored as NCORD = 1, and of ELOAD as NLOAD = 901. This leaves just 900 locations in VECTO allocated to COORD.

This is an excellent solution, which leads to surprisingly efficient and concise coding. It has but one over-riding disadvantage in an educational book, that the names like COORD, ELOAD etc. never appear, so that the coding lacks transparency.

11.14 External attached stiffnesses

The use of prescribed external stiffnesses in finite element analysis is sometimes desirable. For example, it is a particularly convenient way of including cable systems in pressure vessels and other structures. The use of additional stiffnesses avoids the need to model such components by means of specially developed elements. By associating these stiffnesses with the nodal points of an element they can be directly assembled together with the element stiffnesses into the global submatrix, GSTIF. The inclusion of this facility is an almost trivial task which can be undertaken by the reader.

11.15 Stress evaluation—smoothing techniques

Finite element analysis generally involves the minimisation of some functional defined in terms of piecewise functions. These functions are generally required to have a certain degree of inter-element continuity depending on terms in the functional. In many finite element problems, the quantities of primary engineering interest involve the function derivatives and in many instances, especially with lower order elements, these derivatives do not possess inter-element continuity. The analyst is therefore faced with the problem of interpreting quantities which have histogram-type distributions. Often the subjective eye of the experienced analyst may be quite successful in interpreting such information. Equally, it may easily be prejudiced and such an interpretation may lack consistency and rationality. In many of the automatic finite element based systems for linear and non-linear analysis, design and optimisation currently under development, it is crucial that some rational and consistent procedures for the interpretation of discontinuous functions be adopted.

In the displacement method, the stresses are discontinuous between elements because of the nature of the assumed displacement variation. A

typical stress distribution is shown in Fig. 11.3. In analysis involving numerically integrated elements such as isoparametric elements, experience has
shown that the integration points are the best stress sampling points. The
nodes, which are the most useful output locations for stresses, appear to
be the worst sampling points. Reasons for this phenomenon are not immediately apparent. However, it is well known that interpolation functions
tend to behave badly near the extremities of the interpolation region. It is
therefore reasonable to expect that shape function derivatives (and hence

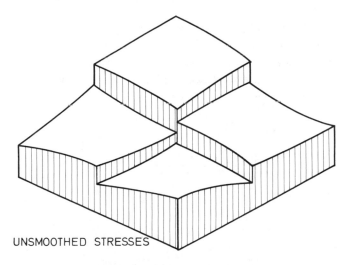

UNSMOOTHED STRESSES

Fig. 11.3.   Unsmoothed stress distribution.

stresses) sampled in the interior of elements will be more accurate than those
sampled on the element periphery. To counteract such problems, many
analysts have taken a nodal average of stresses—average of the nodal stresses
of all elements meeting at a common node. This economic and simple solution works very well on the whole but pays no attention to the size of adjacent
elements.

More specifically, Barlow [9] has shown that for two-dimensional parabolic isoparametric elements the 2×2 Gaussian integrating points are the
optimal stress sampling points. Furthermore, Hinton and Campbell [10]
have shown that a technique known as 'local stress smoothing' is the natural
method for sampling finite element stresses in analyses using reduced integration (i.e. 2×2 Gaussian integration). In this context, local smoothing is
simply a bilinear extrapolation of the 2×2 Gauss point stress values. Thus
smoothed corner node values may be obtained from the Gauss point values

using the expression

$$
\begin{bmatrix} \tilde{\sigma}_1 \\[2ex] \tilde{\sigma}_2 \\[2ex] \tilde{\sigma}_3 \\[2ex] \tilde{\sigma}_4 \end{bmatrix}
=
\begin{bmatrix}
1+\dfrac{\sqrt{3}}{2} & -\dfrac{1}{2} & 1-\dfrac{\sqrt{3}}{2} & -\dfrac{1}{2} \\[2ex]
-\dfrac{1}{2} & 1+\dfrac{\sqrt{3}}{2} & -\dfrac{1}{2} & 1-\dfrac{\sqrt{3}}{2} \\[2ex]
1-\dfrac{\sqrt{3}}{2} & -\dfrac{1}{2} & 1+\dfrac{\sqrt{3}}{2} & -\dfrac{1}{2} \\[2ex]
-\dfrac{1}{2} & 1-\dfrac{\sqrt{3}}{2} & -\dfrac{1}{2} & 1+\dfrac{\sqrt{3}}{2}
\end{bmatrix}
\begin{bmatrix} \sigma_{\mathrm{I}} \\[2ex] \sigma_{\mathrm{II}} \\[2ex] \sigma_{\mathrm{III}} \\[2ex] \sigma_{\mathrm{IV}} \end{bmatrix}
$$

where $\tilde{\sigma}_1$, $\tilde{\sigma}_2$, $\tilde{\sigma}_3$ and $\tilde{\sigma}_4$ are the smoothed corner node values and σ_{I}, σ_{II}, σ_{III} and σ_{IV} are the unsmoothed stresses at the Gauss points as defined in Fig. 11.4.

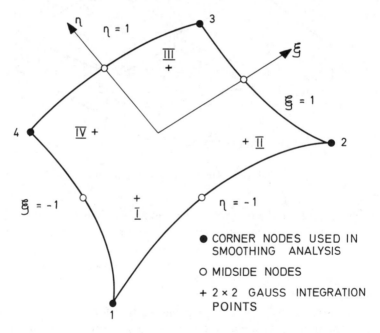

Fig. 11.4.   A two-dimensional parabolic isoparametric element.

This method may easily be incorporated in the stress routines presented earlier and provides a simple method of obtaining nodal stresses. It should be noted that midside node values may be obtained by averaging the values at the nodes associated with the side in question, since the distribution of the smoothed stresses is linear along the sides of the element. Smoothed values should subsequently be averaged to obtain unique values at nodal points.

Further smoothing techniques are described by Oden and fellow workers [11, 12].

11.16 Program structure

Input and output aspects have already been discussed in some detail in Chapter 3 and will not be considered further here. Similarly the advantages of the frontal technique have been adequately covered in Chapter 8. It only remains to discuss briefly the structure of programs when the frontal process is to be employed in solution.

In this text a modular approach has been adopted. This method was chosen so that the various operations of the finite element method could be considered individually and discussed separately. However the very nature of the frontal technique suggests a different program philosophy. Since the main feature of the frontal method is the assimilation and reduction of the stiffness and nodal forces associated with individual elements, economies in core storage and computer costs can be achieved by allowing the frontal solution routine to be the main or master subroutine and performing all operations within an outer element loop. The element stiffnesses and nodal forces would then be computed for one element only, after which its assembly and elimination would take place; the procedure being then repeated for all elements. Such a program structure is particularly beneficial in three-dimensional applications where restrictions on core storage are severe.

11.17 Further application of the programs

The software presented in this text has been employed by the authors as a basis for further research and development programs. Some of these additional applications are listed below.

Linear elastic problems

(i) The comparison of various finite strip methods for plates [13] and box girders [14].
(ii) The solution of some torsion problems [15].
(iii) The comparison of various isoparametric finite elements for the static and dynamic analysis of thick and thin plates [16].
(iv) The analysis of branched shells using 4-noded and Semiloof quadrilateral elements [17], [18].

(v) The transient dynamic analysis of axisymmetric circular plates [19].

Elasto-plastic and elasto-viscoplastic problems

(i) The elasto-plastic and elasto-viscoplastic analysis of plates [20].
(ii) The use of overlay models for elasto-plastic problems [21].
(iii) The analysis of three-dimensional elasto-plastic problems [22].
(iv) The elasto-viscoplastic analysis of 2D solids [23].

Nonlinear quasi-harmonic problems

(i) The analysis of engineering systems governed by the nonlinear quasi-harmonic equation [7].

Nonlinear dynamic transient problems

(i) The transient dynamic analysis of 2D and 3D structures including plasticity, and large deformation effects [24].
(ii) Analysis of thermal shock problems allowing for elasto-plastic material behaviour [25].
(iii) The transient dynamic, large deflection and elasto-plastic analysis of plates [26].
(iv) The transient dynamic, viscoplastic analysis of 2D solids and plates [27].
(v) Transient dynamic crack propagation in 2D and 3D solids [28].

References

1. Grafton, P. E., and Strome, D. R., Analysis of axisymmetric shells by the direct stiffness method. *AIAA J.* **1**, 2342–2347, 1963.
2. Cheung, Y. K., Finite strip method of analysis of elastic slabs. *Proc. Am. Soc. Civ. Eng.* **94**, EM6, 1365–1378, 1968.
3. Gupta, A. K., and Mohraz, B., A method of computing numerically integrated stiffness matrices, *Int. J. Num. Meth. Eng.* **5**, 83–89, 1972.
4. Irons, B. M., The Semiloof shell element, *In* "Finite Elements for Thin Shells and Curved Members," R. H. Gallagher and D. G. Ashwell (Eds), John Wiley & Sons, 1976.
5. Zienkiewicz, O. C., Owen, D. R. J., Phillips, D. V., and Nayak, G. C., Finite element methods in the analysis of reactor vessels. *Nuclear Eng. and Design* **20**, 507–541,1972.
6. Zienkiewicz, O. C., Valliappan, S., and King, I. P., Elasto-plastic solutions of engineering problems. Initial stress, finite element approach. *Int. J. Num. Meth. Eng.* **1**, 75–100, 1969.
7. Lyness, J. F., Owen, D. R. J. and Zienkiewicz, O. C., The finite element analysis of engineering systems governed by a non-linear quasi-harmonic equation *Int. J. Comp. & Struct.* **8**, 65–79, 1975.
8. Zienkiewicz, O. C., "The finite element method in engineering science". 2nd Ed. McGraw-Hill, 1971.

9. Barlow, J., Optimal stress locations in finite element modes. To be published in *Int. J. Num. Meth. Eng.*

10. Hinton, E., and Campbell, J. S., Local and global smoothing of discontinuous finite element function using a least squares method. *Int. J. Num. Meth. Eng.* **8**, 461–480, 1974.

11. Oden, J. T., "Finite Elements of Nonlinear Continua". McGraw-Hill, New York, 1971.

12. Oden, J. T., and Reddy, J. N., Note on an approximate method for computing consistent conjugate stresses in elastic finite elements. *Int. J. Num. Meth. Eng.* **6**, 55–61, 1973.

13. Benson, P. R. and Hinton, E., A thick finite strip solution for static, free vibration and stability problems. *Int. J. Num. Meth. Eng.* **10**, 665–678, 1976.

14. Onate, E., "A comparison of finite strip methods for the analysis of box girder bridges." M.Sc. Thesis, Department of Civil Engineering, University College of Swansea, U.K. 1976.

15. Owen, D. R. J. and Zienkiewicz, O. C., Torsion of axisymmetric solids of variable diameter—including acceleration effects. *Int. J. Num. Meth. Eng*, **8**, 195–198, 1974.

16. Pugh, E. D. L., The static and dynamic analysis of Mindlin plates by isoparametric finite elements. M.Sc. Thesis, Department of Civil Engineering, University College of Swansea, U.K. 1976.

17. Hamill, B. W., The finite element analysis of box girder bridges. M.Sc. Thesis, Department of Civil Engineering, University College of Swansea, U.K., 1976.

18. Martins, R. A. F. and Owen, D. R. J., Structural instability and natural vibration analysis of thin arbitrary shells by use of the Semiloof element. *Int. J. Num. Meth. Eng.* **11**, No. 3, 481–498, 1977.

19. Hinton, E., The dynamic transient analysis of axisymmetric circular plates by the finite element method. *J. Sound Vibs*, **46**, 465–472, 1976.

20. Dinis, L. M. S., Finite element viscoplastic analysis of plates. M.Sc. Thesis, Department of Civil Engineering, University College of Swansea, U.K. 1975.

21. Owen, D. R. J., Prakash, A. and Zienkiewicz, O. C., Finite element analysis of nonlinear composite materials by use of overlay systems. *Int. J. Comp. and Struct.* **4**, 1251–1267, 1974.

22. Owen, D. R. J. and Salonen, E. M., Three dimensional elasto-plastic finite element analysis. *Int. J. Num. Meth. Eng*, **9**, 209–218, 1975.

23. Zienkiewicz, O. C., Owen, D. R. J. and Cormeau, I. C., Analysis of viscoplastic effects in pressure vessels by the finite element method. *Nucl. Eng. and Design* **28**, 278–288, 1974.

24. Shantaram, D., Owen, D. R. J. and Zienkiewicz, O. C., Dynamic transient behaviour of two- and three-dimensional structures including plasticity, large deformation effects and fluid interaction. *Earthquake Eng. and Struct. Dynamics* **4**, 561–578, 1976.

25. Owen, D. R. J. and Selvaraj, S., Thermal shock problems in elasto-plastic solids. Internal Report, Department of Civil Engineering, University College of Swansea, U.K., 1976.

26. Hinton, E., Owen, D. R. J. and Shantaram, D. Dynamic transient nonlinear behaviour of thick and thin plates. *In* "The Mathematics of Finite Elements and Applications, II", MAFELAP 1975, J. R. Whiteman, (ed.) pp. 423–438. Academic Press, London 1977.

27. Owen, D. R. J., Hinton, E. and Shantaram, D. Nonlinear dynamic transient

analysis of plates using parabolic isoparametric elements. Int. Conf. on Finite Elements in Engineering, University of Adelaide, Australia, 1976, Proceedings.
28. Owen, D. R. J. and Shantaram, D., Dynamic crack propagation in two and three dimensional solids by use of finite element methods. EUROMECH Colloquium No. 77, Ecole Polytechnique, Paris, September, 1976.

Appendix I

Instructions for Preparing Input Data

Problems associated with data input have been discussed in detail in Chapter 3 and only users instructions are provided here.

A.1 Beam program, BEAM

CARD SET 1 PROBLEM CARD (I5)—One card
| Cols. 1–5 | NPROB | Total number of problems to be solved in one run |

CARD SET 2 TITLE CARD (12A6)—One card
| Cols. 1–72 | TITLE | Title of the problem—limited to 72 alphanumeric characters |

CARD SET 3 CONTROL CARD (12I5)—One card
Cols. 1–5	NPOIN	Total number of nodal points
6–10	NELEM	Total number of elements
11–15	NVFIX	Total number of restrained boundary points—where one or more degrees of freedom are restrained
16–20	NCASE	Total number of load cases to be analysed
21–25	NTYPE	Blank
26–30	NNODE	Number of nodes per element ($=3$)
31–35	NDOFN	Number of degrees of freedom per node ($=2$)

36–40 NMATS   Total number of different materials
41–45 NPROP   Number of independent properties
                                 per material ($=3$)
46–50 NGAUS   Order of integration formula for
                                 numerical integration (generally use
                                 3)
51–55 NDIME   Number of coordinate dimensions
                                 ($=1$)
56–60 NSTRE   Number of independent generalised
                                 stress components ($=2$)

CARD SET 4 ELEMENT CARDS (4I5)—One card for each element. Total
       of NELEM cards (see Card Set 3)
Cols. 1–5   NUMEL   Element number
       6–10 MATNO(NUMEL)   Material property number
       11–15 LNODS(NUMEL,1)   1st Nodal connection number
       16–20 LNODS(NUMEL,2)   2nd Nodal connection number

CARD SET 5 NODE CARDS (I5, F10.5)—One card for each nodal point
       Total of NPOIN cards (see Card Set 3)
Cols. 1–5   IPOIN   Nodal point number
       6–15 COORD(IPOIN,1)   The x coordinate of the node

CARD SET 6 RESTRAINED NODE CARDS (1X,I4,3X,2I1,2F10.6)—One
       card for each restrained node. Total of NVFIX cards (see
       Card Set 3)
Cols. 2–5   NOFIX(IVFIX)   Restrained node number
       9   IFPRE(IVFIX,1)   Condition of restraint on nodal
                                 displacement, w
                                 $\begin{cases}0 & \text{No displacement restraint}\\1 & \text{Nodal displacement restrained}\end{cases}$
       10   IFPRE(IVFIX,2)   Condition of restraint on nodal
                                 rotation, θ
                                 $\begin{cases}0 & \text{No rotation restraint}\\1 & \text{Nodal rotation restrained}\end{cases}$
       11–20 PRESC(IVFIX,1)   The prescribed value of nodal dis-
                                 placement, w
       21–30 PRESC(IVFIX,2)   The prescribed value of nodal rota-
                                 tion, θ

CARD SET 7 MATERIAL CARDS (I5,3F10.5)—One card for each dif-
       ferent material. Total of NMATS cards (see Card Set 3)

Cols.	1–5	NUMAT	Material identification number
	6–15	PROPS(NUMAT,1)	Flexural rigidity, EI
	16–25	PROPS(NUMAT,2)	Shear constant, GA/1.2
	26–35	PROPS(NUMAT,3)	Distributed load intensity

CARD SETS 2 TO 7 TO BE REPEATED FOR EACH PROBLEM IN ACCORDANCE WITH NPROB IN CARD SET 1.

A.2 The plane stress/strain program, PLANE

CARD SET 1 PROBLEM CARD (I5)—One card

| Cols. | 1–5 | NPROB | Total number of problems to be solved in one run |

CARD SET 2 TITLE CARD (12A6)—One card

| Cols. | 1–72 | TITLE | Title of the problem—limited to 72 alphanumeric characters |

CARD SET 3 CONTROL CARD (12I5)—One card

Cols.	1–5	NPOIN	Total number of nodal points
	6–10	NELEM	Total number of elements
	11–15	NVFIX	Total number of restrained boundary points—where one or more degrees of freedom are restrained
	16–20	NCASE	Total number of load cases to be analysed
	21–25	NTYPE	Problem type parameter $\begin{cases} 1 & \text{Plane Stress} \\ 2 & \text{Plane Strain} \end{cases}$
	26–30	NNODE	Number of nodes per element ($=8$)
	31–35	NDOFN	Number of degrees of freedom per node ($=2$)
	36–40	NMATS	Total number of different materials
	41–45	NPROP	Number of independent properties per material ($=5$)
	46–50	NGAUS	Order of integration formula for numerical integration (generally use 3)
	51–55	NDIME	Number of coordinate dimensions ($=2$)

56–60 NSTRE   Number of independent stress components ($=3$)

CARD SET 4 ELEMENT CARDS (10I5)—One card for each element. Total of NELEM cards (see Card Set 3)

Cols. 1–5 NUMEL   Element Number
       6–10 MATNO(NUMEL)   Material property number
      11–15 LNODS(NUMEL,1)   1st Nodal connection number
      16–20 LNODS(NUMEL,2)   2nd Nodal connection number
           ⋮   ⋮
      46–50 LNODS(NUMEL,8)   8th Nodal connection number

Note: The nodal connection numbers must be listed in an anticlockwise sequence, starting from any corner node

CARD SET 5 NODE CARDS (I5,2F10.5)—One card for each node whose coordinates are to be input

Cols. 1–5 IPOIN   Nodal point number
       6–15 COORD(IPOIN,1) x-coordinate of node
      16–25 COORD(IPOIN,2) y-coordinate of node

Notes: 1) The coordinates of the highest numbered node must be input, regardless of whether it is a midside node or not.
       2) The total number of cards in this set will generally differ from NPOIN (see Card Set 3) since for element sides which are linear it is only necessary to specify data for corner nodes; intermediate nodal coordinates being automatically interpolated if on a straight line.

CARD SET 6 RESTRAINED NODE CARDS (1X,I4,3X,2I1,2F10.6)—One card for each restrained node. Total of NVFIX cards (see Card Set 3)

Cols. 2–5 NOFIX(IVFIX)   Restrained node number
       9   IFPRE(IVFIX,1)   Condition of restraint on x displacement
                                 $\begin{cases} 0 & \text{No displacement restraint} \\ 1 & \text{Nodal displacement restrained} \end{cases}$
      10   IFPRE(IVFIX,2)   Condition of restraint on y displacement
                                 $\begin{cases} 0 & \text{No displacement restraint} \\ 1 & \text{Nodal displacement restrained} \end{cases}$
      11–20 PRESC(IVFIX,1)   The prescribed value of the x component of nodal displacement ·

21–30 PRESC(IVFIX,2) The prescribed value of the y component of nodal displacement

CARD SET 7 MATERIAL CARDS (I5,5F10.5)—One card for each different material. Total of NMATS cards (see Card Set 3)

Cols.		
1–5	NUMAT	Material identification number
6–15	PROPS(NUMAT,1)	Elastic modulus, E
16–25	PROPS(NUMAT,2)	Poisson's ratio, v
26–35	PROPS(NUMAT,3)	Material thickness, t
36–45	PROPS(NUMAT,4)	Mass density, ρ
46–55	PROPS(NUMAT,5)	Coefficient of thermal expansion, α

CARD SET 8 LOAD CASE TITLE CARD (12A6)—One card
Cols. 1–72 TITLE Title of the load case—limited to 72 alphanumeric characters

CARD SET 9 LOAD CONTROL CARD (4I5)—One card
Cols. 1–5 IPLOD Applied point load control parameter

$\begin{cases} 0 & \text{no applied nodal loads to be input} \\ 1 & \text{applied nodal loads to be input} \end{cases}$

6–10 IGRAV Gravity loading control parameter

$\begin{cases} 0 & \text{no gravity loads to be considered} \\ 1 & \text{gravity loads to be considered} \end{cases}$

11–15 IEDGE Distributed edge load control parameter

$\begin{cases} 0 & \text{no distributed edge loads to be input} \\ 1 & \text{distributed edge loads to be input} \end{cases}$

16–20 ITEMP Thermal loading control parameter

$\begin{cases} 0 & \text{no thermal loading to be considered} \\ 1 & \text{thermal loading to be considered} \end{cases}$

CARD SET 10 APPLIED LOAD CARDS (I5, 2F10.3)—One card for each loaded nodal point

Cols.		
1–5	LODPT	Node number
6–15	POINT(1)	Load component in x direction
16–25	POINT(2)	Load component in y direction

Notes: 1) The last card should be that for the highest numbered node whether it is loaded or not
2) If IPLOD = 0 in Card Set 9, omit this set.

CARD SET 11 GRAVITY LOADING CARD (2F10.3)—One card

Cols.	1–10	THETA	Angle of gravity axis from the positive y axis (see Figure 7.1)
	11–20	GRAVY	Gravity constant—specified as a multiple of the gravitational acceleration, g

Note: If IGRAV = 0 in Card Set 9, omit this set.

CARD SET 12 DISTRIBUTED EDGE LOAD CARDS

12(a) CONTROL CARD (I5)—One card

Cols.	1–5	NEDGE	Number of element edges on which distributed loads are to be applied

12(b) ELEMENT FACE TOPOLOGY CARD (4I5)

Cols.	1–5	NEASS	The element number with which the element edge is associated
	6–10	NOPRS(1)	List of nodal points, in an anti-
	11–15	NOPRS(2)	clockwise sequence, of the nodes
	16–20	NOPRS(3)	forming the element face on which the distributed load acts.

12(c) DISTRIBUTED LOAD CARDS (6F10.3)

Cols.	1–10	PRESS(1,1)	Value of normal component of distributed load at node NOPRS(1)
	11–20	PRESS(2,1)	Value of normal component of distributed load at node NOPRS(2)
	21–30	PRESS(3,1)	Value of normal component of distributed load at node NOPRS(3)
	31–40	PRESS(1,2)	Value of tangential component of distributed load at node NOPRS(1)
	41–50	PRESS(2,2)	Value of tangential component of distributed load at node NOPRS(2)
	51–60	PRESS(3,2)	Value of tangential component of distributed load at node NOPRS(3)

Notes   1) Subsets 12(b) and 12(c) must be repeated in turn for every element edge on which a distributed load acts. The element edges can be considered in any order.

2) If IEDGE = 0 in Card Set 9, omit this card set.

CARD SET 13 TEMPERATURE CARDS (I5,F10.3)
Cols. 1–5   NODPT   Node number
       6–15   TEMPE(NODPT)   Temperature at node

Notes 1) Datum temperature is taken to be zero
       2) Only nodal temperatures which are non-zero need be input.
          The card set must terminate with the highest numbered node
          regardless of the temperature value at this node
       3) If ITEMP = 0 in Card Set 9, omit this set.

CARD SETS 8 TO 13 TO BE REPEATED FOR EACH LOAD CASE IN
ACCORDANCE WITH NCASE IN CARD SET 3.

CARD SETS 2 TO 13 TO BE REPEATED FOR EACH PROBLEM IN
ACCORDANCE WITH NPROB IN CARD SET 1.

A.3 The plate bending program, PLATE

CARD SET 1 PROBLEM CARD (I5)—One card
Cols. 1–5   NPROB   Total number of problems to be
                                       solved in one run

CARD SET 2 TITLE CARD (12A6)—One card
Cols. 1–72   TITLE   Title of the problem—limited to 72
                                       alphanumeric characters

CARD SET 3 CONTROL CARD (12I5)—One card
Cols. 1–5   NPOIN   Total number of nodal points
       6–10   NELEM   Total number of elements
       11–15 NVFIX   Total number of restrained bound-
                                       ary points—where one or more
                                       degrees of freedom are restrained
       16–20 NCASE   Total number of load cases to be
                                       analysed
       21–25 NTYPE   Blank
       26–30 NNODE   Number of nodes per element $(=8)$
       31–35 NDOFN   Number of degrees of freedom per
                                       node $(=3)$
       37–40 NMATS   Total number of different materials

41–45	NPROP	Number of independent properties per material $(=4)$
46–50	NGAUS	Order of integration formula for numerical integration (generally use 2)
51–55	NDIME	Number of coordinate dimensions $(=2)$
56–60	NSTRE	Number of independent generalised stress components $(=5)$

CARD SET 4 ELEMENT CARDS (10I5)—One card for each element. Total of NELEM cards (see Card Set 3)

Cols.	1–5	NUMEL	Element number
	6–10	MATNO(NUMEL)	Material property number
	11–15	LNODS(NUMEL,1)	1st Nodal connection number
	16–20	LNODS(NUMEL,2)	2nd Nodal connection number
		$\vdots$	$\vdots$
	46–50	LNODS(NUMEL,8)	8th Nodal connection number

Note: The nodal connection numbers must be listed in an anticlockwise sequence, starting from any corner node; the element being viewed from above the plane $z = 0$. (See Fig. 3.3)

CARD SET 5 NODE CARDS (I5,2F10.5)—One card for each node whose coordinates are to be input

Cols.	1–5	IPOIN	Nodal point number
	6–15	COORD(IPOIN,1)	x-coordinate of node
	16–25	COORD(IPOIN,2)	y-coordinate of node

Notes 1) The coordinates of the highest numbered node must be input, regardless of whether it is a midside node or not.

2) The total number of cards in this set will generally differ from NPOIN (see Card Set 3) since for element sides which are linear it is only necessary to specify data for corner nodes; intermediate nodal coordinates being automatically interpolated if on a straight line.

CARD SET 6 RESTRAINED NODE CARDS (1X,I4,2X,3I1,2F10.6)— One card for each restrained node. Total of NVFIX cards (see Card Set 3).

| Cols. | 2–5 | NOFIX(IVFIX) | Restrained node number |

8	IFPRE(IVFIX,1)	Condition of restraint on nodal displacement, w

$$\begin{cases} 0 & \text{No displacement restraint} \\ 1 & \text{Nodal displacement restrained} \end{cases}$$

9	IFPRE(IVFIX,2)	Condition of restraint on nodal rotation, θ_x

$$\begin{cases} 0 & \text{No rotation restraint} \\ 1 & \text{Nodal rotation restrained} \end{cases}$$

10	IFPRE(IVFIX,3)	Condition of restraint on nodal rotation, θ_y

$$\begin{cases} 0 & \text{No rotation restraint} \\ 1 & \text{Nodal rotation restrained} \end{cases}$$

11–20	PRESC(IVFIX,1)	The prescribed value of nodal displacement, w
21–30	PRESC(IVFIX,2)	The prescribed value of nodal rotation, θ_x
31–40	PRESC(IVFIX,3)	The prescribed value of nodal rotation, θ_y

CARD SET 7 MATERIAL CARDS (I5,4F10.5)—One card for each different material. Total of NMATS cards (see Card Set 3)

Cols.	1–5	NUMAT	Material identification number
	6–15	PROPS(NUMAT,1)	Elastic modulus, E
	16–25	PROPS(NUMAT,2)	Poisson's ratio, v
	26–35	PROPS(NUMAT,3)	Material thickness, t
	36–45	PROPS(NUMAT,4)	Intensity of any uniformly distributed load

CARD SET 8 LOAD CASE TITLE CARD (12A6)—One card

Cols.	1–72	TITLE	Title of the load case—limited to 72 alphanumeric characters

CARD SET 9 LOAD CONTROL CARD (I5)—One card

Cols.	1–5	IPLOD	Applied point load control parameter

$$\begin{cases} 0 & \text{No applied nodal loads to be input} \\ 1 & \text{Applied nodal loads to be input} \end{cases}$$

CARD SET 10 APPLIED LOAD CARDS (I5,3F10.3)—One card for each loaded nodal point

Cols.	1–5	LODPT	Node number
	6–15	POINT(1)	Load component in the z direction
	16–25	POINT(2)	Nodal couple in the xz plane
	26–35	POINT(3)	Nodal couple in the yz plane

Notes 1) The last card should be that for the highest numbered node whether it is loaded or not

2) If IPLOD = 0 in Card Set 9, omit this set.

CARD SETS 8 TO 10 TO BE REPEATED FOR EACH LOAD CASE IN ACCORDANCE WITH NCASE IN CARD SET 3.

CARD SETS 2 TO 10 TO BE REPEATED FOR EACH PROBLEM IN ACCORDANCE WITH NPROB IN CARD SET 1.

Appendix II

Dictionary of Variable Names

ASDIS(MSVAB)

vector of nodal displacements calculated using FRONT—the sequence is $(w_1, \theta_1, w_2, \theta_2, w_3, \theta_3;$ etc.) for the beam, etc.

BMATX(NSTRE,NEVAB)

the element strain matrix at any point within the element, e.g. for a beam element

$$\mathbf{B} = [\mathbf{B}_1, \mathbf{B}_2, \mathbf{B}_3]$$

CARTD(NDIME,NNODE)

Cartesian shape function derivatives associated with nodes of current element. e.g. for a beam element

$$\left[\frac{\partial N_1}{\partial x}, \frac{\partial N_2}{\partial x}, \frac{\partial N_3}{\partial x} \right]$$

COORD(NPOIN,NDIME)

coordinates of nodal points

DAREA

an infinitesimal element of area at a Gauss point × the Gaussian weight coefficients (used in plate bending program)

DBMAT(NSTRE,NEVAB)

the result of the matrix multiplication **DB**

DERIV(NDIME,NNODE)

shape function derivatives associated with nodes of the current element—

295

sampled at any point within the element, e.g. for a beam element

$$\left[\frac{\partial N_1}{\partial \xi}, \frac{\partial N_2}{\partial \xi}, \frac{\partial N_3}{\partial \xi}\right]$$

DJACB
determinant of the Jacobian matrix sampled at any point within the element

DLENG
an infinitesimal element of length at a Gauss point × the Gaussian weight coefficient (used in beam element program)

DMATX(NSTRE,NSTRE)
the matrix of elastic constants or rigidities **D**

DVOLU
an infinitesimal element of volume at a Gauss point × the Gaussian weight coefficients (used in plane stress/strain program)

ELCOD(NDIME,NNODE)
local array of nodal Cartesian co-ordinates for the element currently under consideration

ELDIS(NDOFN,NNODE)
nodal displacements associated with a particular element—to be used in stress resultant calculations

ELOAD(NELEM,NEVAB)
nodal forces for each element

EQRHS
The right hand side load term after reduction, ready for disc file storage—used in FRONT

EQUAT(IFRON)
The array in which the reduced equations are stored prior to writing to disc in FRONT

ESTIF(NEVAB,NEVAB)
the element stiffness matrix $\mathbf{K}^e$

ETASP
η coordinate of a sampling point—this is usually a Gauss point

N.B. ETASP = T in subroutine SFR2

EXISP	ξ coordinate of a sampling point— this is usually a Gauss point N.B. EXISP = S in subroutines SFR1 and SFR2
FIXED(IPOSN)	Prescribed displacement values transferred from PRESC array in FRONT. (See also IFFIX array)
GPCOD(NDIME,NGASP)	local array of Cartesian coordinates of Gauss points for element currently under consideration
GSTIF(ISTIF)	global stiffness matrix used in FRONT N.B. Only upper triangle is stored in vector form
IFFIX(IPOSN)	Fixity integers transferred from IFPRE array in FRONT. Index IPOSN determines the position of a particular nodal degree of freedom in the vector. The IDOFNth degree of freedom of node IPOIN is located according to IPOSN = (IPOIN − 1) *NDOFN + IDOFN. The array is first zeroed with the unit node values being then inserted in the current locations
IFPRE(IVFIX,IDOFN)	integer code to specify which degrees of freedom at a node are to be restrained or prescribed with specified displacement values, e.g. for plane stress/strain 10 Displacement in the x direction restrained 01 Displacement in the y direction restrained 11 Displacement in the x and y directions restrained (Only applicable for boundary nodes)
KGASP,NGASP	Kounter, number of Gauss points used. NGASP= 2 for a beam element

LNODE	node currently under consideration
LNODS(NELEM,NNODE)	element node numbers listed for each element
LOCEL(IEVAB)	the vector which locates the global position of each element variable— used in FRONT
LPROP	material set of element currently under consideration
MATNO(NELEM)	material set numbers for each element
MMATS,NMATS	maximum, number of material sets MMATS is never actually used in the program and is given here simply for convenience, e.g. MMATS = 10 for the beam program.
MPOIN	Maximum number of nodal points MPOIN is never actually used in the program and is given here simply for convenience e.g. MPOIN = 50 for the beam program
MSVAB = NTOTV	Maximum number of structural variables MSVAB = MPOIN*NDOFN. MSVAB is never actually used in the program and is given here simply for convenience. e.g. MSVAB = 100 for the beam program
NACVA(IFRON)	vector of active variables used in FRONT
NCASE	The total number of load cases to be solved for. Provided that the structural geometry remains unchanged the element stiffnesses need not be recomputed for each additional loading case.
NDEST(IEVAB)	Destination vector used in FRONT
NDIME	Number of coordinate components required to define each nodal point

	1 Beam analysis 2 Plane stress/strain 2 Plate bending
NDOFN	The number of degrees of freedom per nodal point. For beam analysis this is $2(w, \theta)$, for two dimensional plane solids it is also 2 (u, v) and for plate bending problems it is 3 (w, θ_x, θ_y)
NELEM	Total number of elements in the structure
NEVAB	Number of variables per element NEVAB = 6 for a beam element
NGAST	Total number of Gauss points in a finite element mesh
NGAUS	Number of Gauss rule adopted
NIKNO	Nickname number for a particular variable used in FRONT
NNODE	Number of nodes per element 3 for the beam program, 8 for plane stress/strain, 8 for plate bending
NOFIX(IVFIX)	signifies that the IVFIXth boundary node to be specified has a nodal point number NOFIX(IVFIX)
NOPRS(IODEG)	defines edge of plane stress/strain element on which a distributed load is acting
NPOIN	Total number of nodal points in the structure
NPROP	The number of material parameters required to define the characteristics of a material completely. 3 Beam analysis 5 Plane Stress/strain 4 Plate bending These are discussed in detail in Section 3.6

NSTRE   Number of stress components at any point
                               2 Beam program
                               3 Plane stress/strain
                               5 Plate bending

NTOTV   number of total variables in the structure

NTYPE   Problem type $\begin{cases} 1\text{--Plane stress} \\ 2\text{--Plane strain} \end{cases}$
                               This parameter is not applicable for beam and plate bending applications. For these cases enter a zero value

NVFIX   Total number of boundary points, i.e. nodal points at which one or more degrees of freedom are restrained. It should be noted that in this context an internal node can be a boundary node.

PIVOT   The diagonal pivoting term used in FRONT

POINT(IDOFN)   Applied nodal point forces

POSGP(NGAUS) $\left. \begin{matrix} \xi \\ \eta \end{matrix} \right\}$ coordinates of Gauss points

PRESC(IVFIX,IDOFN)   indicates that the IDOFNth degree of freedom of the IVFIXth boundary node has a prescribed value provided that the corresponding value of IFPRE(IVFIX,IDOFN) is equal to 1.

PRESS(IODEG,IDOFN)   Values of the normal and tangential load intensities at nodes of an edge of a plane stress/strain element

PROPS(NMATS,NPROP)   Material properties for each material set

SHAPE(NNODE)   Shape functions associated with each node of the current element sampled at:

(i) for the beam element, any point ξ_P within the element

$$[N_1(\xi_P), N_2(\xi_P), N_3(\xi_P)]^T$$

(ii) for the plane stress/strain and plate bending elements any point (ξ_P, η_P) within the element

$$[N_1(\xi_P, \eta_P), N_2(\xi_P, \eta_P), \ldots,$$

$$N_8(\xi_P, \eta_P)]^T$$

SMATX(NSTRE,NEVAB,NGASP)   contains DBMAT for each Gauss point—element stress matrix

STRIN(JSTRE,KGAST)   initial stresses

$$\boldsymbol{\sigma}^0 = [\sigma_x^0, \sigma_y^0, \tau_{xy}^0, \sigma_z^0]^T$$

at each Gauss point for plane strain application

STRSG(NSTRE)

(i) for the beam element stress resultants at Gauss point ξ_P for current element, i.e.

$$[M(\xi_P), Q(\xi_P)]^T$$

(ii) for plane stress/strain element, stresses at Gauss point (ξ_P, η_P) for current element   i.e.

$$[\sigma_x(\xi_P, \eta_P), \sigma_y(\xi_P, \eta_P), \tau_{xy}(\xi_P, \eta_P),$$

$$\sigma_z(\xi_P, \eta_P)]^T$$

(iii) for plate bending element, stress resultants at Gauss point (ξ_P, η_P) for current element, i.e.

$$[M_x(\xi_P, \eta_P), M_y(\xi_P, \eta_P),$$

$$M_{xy}(\xi_P, \eta_P), Q_x(\xi_P, \eta_P),$$

$$Q_y(\xi_P, \eta_P)]^T$$

STRSP(NSTRE)   Maximum principal stress, minimum principal stress and angle α which the maximum principal stress makes with the positive x axis. Used in plane stress/strain program

VECRV(IFRON)   The vector of running variables in
                                     which the solved displacements are
                                     stored in FRONT

WEIGP(NGAUS)   weighting factors for Gauss points

INDEX